W9-CSY-571

IN BAD TASTE

The MSG Syndrome

How living without MSG can reduce headache, depression and asthma, and help you get control of your life

George R. Schwartz, M.D.

Foreword by
Arthur D. Colman, M.D.

HEALTHPRESS
P.O. Box 367, Santa Fe, New Mexico 87501
Santa Fe • San Francisco

Published by
Health Press
A Division of Healing Research, Inc.
P.O. Box 367
Santa Fe, New Mexico 87501

ISBN 0-929173-00-7
Library of Congress Card Catalog Number 88-80783
Printed in the United States of America

Designed and produced by Linda Seals
10 9 8 7 6 5 4 3

Editor: Harriet Slavitz

CONTENTS

Foreword by Arthur D. Colman, M.D.

PART I

Chapter 1	Introduction	1
Chapter 2	MSG History	5
Chapter 3	No Laughing Matter *Case Histories (including migraine, arthritis-like symptoms, dizziness, slurred speech and others)* *MSG Intolerance in Children* *Asthma and MSG* *Depression and MSG* *Headaches and MSG* *Aging and MSG* *MSG and Women's Menstrual Cycles*	11
Chapter 4	The MSG Syndrome: What You Don't Know Can Hurt You	35
Chapter 5	MSG Symptom Analysis *Are You Reacting to MSG?* *Mechanisms of MSG Toxicity*	39

Chapter 6	MSG: Bad Taste *Glutamate as a Neurotransmitter* *Animal Studies of MSG*	43
PART II		
Chapter 7	Living Without MSG: MSG-Free Cooking at Home *Easy Food Shopping Tips* *Substitutions*	47
Chapter 8	Shopping Smart and Reading Labels *MSG and Food Labelling* *MSG and the Supermarket*	51
Chapter 9	Restaurant Guide to Food Containing MSG *Fast Food Restaurants* *Restaurants and Cafeterias* *Airplanes and Trains*	81
Chapter 10	Conclusion *MSG as a Drug*	89
Acknowledgments		93
Appendix 1	Recipes	97
Appendix 2	Account of Our First Observations of MSG Asthma (Dr. Gary Baker)	109
Index		113

Dedication
To all the MSG-sensitive people with the hope that this book will help to identify and relieve their distress.

It is dangerous to be right in matters on which the established authorities are wrong.

Voltaire

FOREWORD

Is MSG dangerous to your health? How can you tell for yourself? And if your body is reactive to this substance, what steps would you need to take to eliminate MSG, as well as the prepared foods and additives that contain it, from your diet? At last we have a book available which provides sufficient historical, scientific, clinical, and nutritional data to decide for yourself about the potential dangers of MSG to you and your children.

The publication of this book, the first of its kind, aimed at both a professional and non-professional audience, is a significant event for a large and growing segment of the population—estimated at more than 25%—who react to MSG. Many suffer in ignorance despite the fact that it has been more than 20 years since physicians and scientists began publishing well-documented accounts of serious, even life-threatening, effects from the ingestion of small amounts of MSG found in ordinary canned soup, fast food, or seasoning salts. New articles continue to appear regularly in the widely circulated scientific journals, detailing the deleterious effects of MSG toxicity. But there has been a lack of accurate and educational information necessary for the public to judge the effects of MSG on its own.

I have become increasingly aware of how many of my patients and associates developed the well-documented headaches, facial flush, gastrointestinal symptoms, and depression. In 1978 I published an account in the *New England Journal of Medicine* which detailed the case histories of two individuals who developed physical and psychiatric symptoms, including severe depression, whenever they ate foods containing MSG. Their emotional symptoms were particularly disturbing because they began as late as 48 hours after the exposure to MSG and sometimes continued for several weeks. None of the symptoms were present as long as MSG was removed from their diet.

One of the individuals was a child whose school and home adjustment had been severely affected for years by the MSG present in his very ordinary diet. Completely eliminating MSG improved his behavior to such a degree that within weeks his teacher and psychiatrist were calling his parents to ask what had changed! Ten years later this boy, once considered "hyperactive and difficult" is an honor chemistry student at the University of California at Berkeley.

The *New England Journal of Medicine* article provoked instant publicity and extensive press coverage. I reviewed hundreds of letters from doctors and other professionals providing their own clinical experiences, as well as correspondence from MSG sufferers who detailed their difficulties.

Other responses were less professional. The leading Japanese newspaper in a West Coast metropolis scheduled and then cancelled an interview. They were "being threatened with loss of large food advertisers." The *New England Journal of Medicine* received an antagonistic letter from the "Glutamate Association," a multi-million-dollar worldwide lobbying organization, stating that they were "disappointed and very concerned" about the publication of my letter. They imputed the standards of the *Journal* for printing it, suggesting that "a formal statement of policy or communication to the media" be made by the *Journal* as a statement of its criteria for publication.

The campaign against my three-paragraph account was not yet over. Another letter was received by the *Journal* from a researcher who questioned my clinical findings. I received space from the editors to refute the arguments raised. The task was not difficult, since his letter contained more propaganda than science.

Eventually the letters to the *Journal* stopped, but the harrassment was not over. A classmate of mine from Harvard Medical School, whom I had not seen for over ten years, suddenly appeared in San Francisco. We met for breakfast and our conversation quickly turned to my work on the psychiatric syndrome associated with MSG. He began by disparaging my findings, and then offered me funding to do research on other "allergic phenomena" instead. I was less bothered by his aspersions on my clinical descriptions than by his motivation for doing so. When I pressed him about his special interest he admitted to being a paid consultant for the Glutamate Foundation. He became angry when I called him a "hit man" and I have not heard from my Harvard "friend" since.

Obviously, there is an industry at stake in protecting the reputation of MSG. Yet, despite the fortune spent on advertising and public relations which laud its "natural" taste-enhancing effects, a wariness of the chemical is growing.

In my clinical practice I routinely ask about possible MSG toxicity whenever there is a combination of gastrointestinal, neurological, and depressive symptoms and sometimes I am able to save the patient a great deal of suffering and expense. I know of pediatricians who ask parents about MSG in the diets of symptomatic children. Some even administer test doses of the chemical in their offices to rule out or confirm their suspicions. Efforts of these kinds are hampered by a too diffuse focus on general "food allergies" rather than specific toxicities. It is important to remember that in sufficient quantities, MSG is toxic to everyone. To those who cannot metabolize it effectively, it acts like a poison.

Unfortunately, under present circumstances elimination of MSG is not a simple matter. A parent or concerned consumer might have a Ph.D. in nutrition and still not be able to completely eliminate MSG.

That's why this book is so valuable. If you or anyone you know is an MSG sufferer, it is hardly an exaggeration to say that this book could change your life.

Arthur D. Colman, M.D.
Clinical Professor
Department of Psychiatry
U.C. Medical Center
and Private Practice
Sausalito, California

CHAPTER ONE

Introduction

After salt and pepper is a third spice, the most widely used flavor enhancer in the world, and truly the spice of our times. In America it is called monosodium glutamate, in Japan Aji-no-moto, and it is known by other names in other countries.

The substance was first developed in 1908 in the laboratory of Kikunae Ikeda as an isolate of a flavor-enhancing seaweed known as kombu or "sea tangle." Dr. Ikeda hardly could have suspected then that his work to identify the active substance in a seaweed which Japanese chefs had used for thousands of years would lead to a multi-billion-dollar industry in the twentieth century.

Shortly after he isolated MSG, Kikunae Ikeda became a partner in what would become the Aji-no-moto (the "essence of taste") Company. In fact, throughout the Orient MSG is known as Aji-no-moto, much as "xerox" is used as a generic term for copying machines or "kleenex" for all paper tissues in the United States. In recognition of the importance of his work, Dr. Ikeda's original isolated substance is encased in a monument at Tokyo University.

Today MSG is used to enliven the taste of processed foods and fast-foods and to give Chinese food its tasty quality. Found in most commercial soups, soy sauce, and the additive hydrolyzed vegetable protein,* MSG has become a staple of the modern food industry.

Why should we be concerned about a substance which can make dreary foods taste better, remove the tinny taste from canned foods, and stimulate the tired taste buds? Why should anyone want to challenge these apparently beneficial properties of MSG?

The simple answer is that MSG is a poison to many people who are sensitive to its effects.

Reactions among the MSG-sensitive range from mild to very severe. Indeed, the symptoms that Dr. Ho Man Kwok reported in the first published study in the *New England Journal of Medicine* in 1968, for example headache and flushing of the skin, were relatively mild. However, later studies have documented more serious and sustained physical problems such as asthma, acute headaches, and life-threatening heart irregularities. Other symptoms that might seem to be psychological in origin have also been traced to MSG consumption: extreme mood swings, irritability, depression, and even paranoia.

Many cases of severe problems induced by MSG have been documented by physicians and more have been reported in the medical literature where they may be studied by professionals. But the vast majority of MSG-sensitive people are not aware of the problems this substance may be creating in their lives: the fruitless visits to physicians who cannot explain their complaints, the anger and depression it can induce, the danger even of suicide if these emotional disturbances are not traced to their source.

A questionnaire study (reported in Federation Proceedings** in April, 1977) using a large sampling of subjects done by Dr. Liane Reif-Lehrer showed that 30% of adults and between ten percent and 20% of children have some reaction to foods con-

*Hydrolyzing vegetable protein is one of the chemical methods of producing MSG. This mixture, containing up to 20% MSG, is listed among the ingredients of many commonly used processed foods.

**The Federation of the American Society for Experimental Biology.

taining MSG. This means that many millions of lives are adversely affected by this substance.

Monosodium glutamate is found in most of the food prepared by major fast-food chains such as Kentucky Fried Chicken. With the popularity of these foods among children and teenagers, it could be a factor in "Agent Blue" to which the National Institute of Mental Health researchers attribute the recent rise in child and teen depressive syndromes. Behavioral and physical problems of children, such as incontinence and seizures, have been diagnosed and successfully treated as MSG reactions.

MSG was removed from baby foods in the late 1960s without much comment. Dr. Jean Mayer, the noted Harvard food scientist, remarked at a women's meeting of the National Press Club that "with even the slightest presumption of guilt I would take the damn stuff out of baby food." Gerber, Heinz, and Beechnut almost immediately announced that they would stop using MSG in baby foods. *However, infants still get MSG from ordinary table foods.*

The body of scientific facts now has reached a level where the findings and case reports must be brought to public attention. At least 20 million people in the United States and more than 100 million people worldwide react to MSG. This means that although some people can ingest MSG with impunity, many others are being damaged physically and emotionally by the unknowing use of this flavor enhancer which for them has potent drug effects.

In Bad Taste: The MSG Syndrome is the story of the adverse effects of monosodium glutamate on the health and well being of some consumers. It details the studies of dedicated scientists who have warned against MSG use. It explores an economy partly dependent on MSG production and use; and how psychologists, physicians, clinics, schools, and lawyers have become involved in the problem due to its widespread effects.

Laboratory and field studies describe a consistent picture: 30% of the population experience some symptoms from MSG in amounts commonly added to foods. Clinical data show that some of these individuals develop symptoms that are not mild or transient, but intense and dangerous; possibly, although less commonly, developing chronic long-term and disabling problems. More and more individuals will be affected as the use of MSG continues to increase. MSG intolerance is not an allergic reaction—but a true drug effect. A high enough dose can affect anyone; and many more people are reaching that symptom-causing dose in MSG consumption. The MSG syndrome is not a bizarre reaction to some ethnic food (i.e., Chinese) as some people still believe.

People who react to MSG must first identify their reactions and then learn to eat food without this additive. This book will serve to guide the consumer through the supermarket aisles on preventive shopping trips and provide tasty recipes which eliminate MSG from the diet. A discussion of restaurants and dining offers tips on avoiding MSG, which may be found even in the finest kitchens. There is also a review of fast-food chain favorites containing MSG.

People who react severely to MSG experience almost continual distressing physical and psychological symptoms. Knowing how to avoid this flavor enhancer can dramatically change lives. In order to unravel the story of MSG, we must start with the beginning: MSG history. (Those readers concerned with MSG reactions may want to go directly to Chapter 3 and read about the history of MSG at a later time.)

CHAPTER TWO

MSG History

Thousands of years ago when the Japanese began using seaweed as a food they discovered that sea tangle or "kombu" (*Laminaria japonica*) was a good seasoning ingredient. At that time the active flavor-enhancing substance within the sea tangle (MSG) had not been identified.

In 1860, in Hamburg, Germany, the chemist Rittenhausen was working on the determination of the chemical make-up of animal proteins, particularly the amino acid, glutamic acid. There is no evidence that Rittenhausen (and later another German chemist, Wolff) had any interest in or knowledge of cooking or flavor-enhancing. They were pure scientists, trying to identify the chemical properties of the various protein substances. Their work was essential, however, to allow Kikunae Ikeda's identification of the active flavor-enhancing substance in seaweed and its eventual manufacture.

Ikeda, a Kyoto boy, was educated at Tokyo University and graduated with a degree in chemistry in 1889. After briefly teaching at a high school, he went to Germany to further his studies. In Germany Ikeda became involved with the important work Dr. Wolff was doing in protein chemistry. Glutamic acid had been synthesized in 1890 and Ikeda learned the chemical techniques of identification and synthesis during his apprenticeship of several years.

Returning to Japan, he took a post in the chemistry department of the Imperial University of Tokyo. By 1908 an established faculty member, he began to investigate the seaweed his wife was using to flavor the family's soup and soon had identified the flavor-enhancing substance.

He discovered that the active extract of the flavorful seaweed had the characteristics of glutamic acid, and that this active substance was monosodium glutamate, the sodium salt of glutamic acid. His discovery was secured with British Patent number 9440—the title, "Manufacture of a Flavoring Material"—on April 21, 1909.

Kikunae Ikeda knew he held a food secret with many possible applications. While still at the Imperial University of Tokyo he began filing other patents on commercial manufacturing processes, first using the breakdown of wheat protein and later soybean protein.

In 1909 he teamed up with another enterprising businessman, former pharmacist Saburosuke Suzuki, and convinced him that they were on the verge of a business which could add a new flavor, Unami, to the world. The Aji-no-moto name was chosen as their trademark and later this became the name of the company responsible for the development and distribution of the flavor enhancer throughout the world. "Aji" means the origin, the beginning, or at the foot of; "moto" translates as taste or flavor. Therefore, Aji-no-moto literally signifies "At the Origin of Flavor."

By 1933 Japanese production of the substance we know as MSG had reached ten million pounds yearly and it had become the most important flavoring in the Orient. Today Aji-no-moto produces more than half of the world's supply of MSG.

In the early days, only once was the domination of the Aji-no-moto Company challenged. In China production of MSG began in the 1920s and reached almost 400,000 pounds annually by 1930. By the mid-1930s the Chinese were offering the Japanese serious competition with their product, marketed under the name Ve-tsin. However, the Japanese invasion of coastal

China in 1937 put the Chinese glutamate factories out of commission.

Aji-no-moto did not find easy acceptance in America, although not for lack of trying, with efforts to gain acceptance begun in the early 1920s. The story of its arrival made little impact and the man who helped to stimulate its development has long been forgotten.

In 1925, James E. Larrowe and the Larrowe Milling Company contacted the Suzuki Spice Company for help in the profitable disposal of "waste water," a residue of the processing of sugar from sugar beets which contains a substantive amount of glutamic acid. Sometimes called "Steffen's waste water" after the man who developed the process, America in 1925 had a surplus of the liquid. Or more accurately, Larrowe Milling Company had a surplus after World War I.

The United States was cut off from its major source of potash (Germany) for fertilizer and it became profitable to extract potassium salts from the waste water. Many sugar beet processing plants were converted into potash plants, spurred by prices of $400 per ton. But in 1918, with the end of the war, the price of potash fell drastically and Larrowe was left with many thousands of tons of "Steffen's waste water" in great storage tanks in Mason City, Iowa. Larrowe had expanded his production to meet an anticipated potash demand and now was in financial straits. He had to find a use for the waste water.

Almost every possible use, including anti-freeze for automobiles, was tried without success. In desperation Larrowe went to the Mellon Institute of Industrial Research in Pittsburgh, Pennsylvania, and presented his problem. By now it was 1923 and the waste water had been sitting for over five years. Millions of dollars were tied up in what seemed to be hopeless material.

After almost ruining the engine of his new Packard, Larrowe wanted the Institute to determine why he couldn't use the material as anti-freeze. Dr. Donald Tressler, a young scientist, was offered a fellowship to analyze and find a use for the waste water. After two years of research he found that the waste water contained appreciable amounts of glutamic acid. Upon consulting with his professor, Dr. Elmer McCollum, a well-known vitamin chemist at Johns Hopkins University, he was advised that Larrowe should drop the idea of anti-freeze and set out to manufacture glutamate.

Larrowe contacted Suzuki and Ikeda and offered them an opportunity to buy his waste water glutamate and in 1926 Dr. Ideka, Mr. Suzuki, and Mr. Suzuki's son arrived from Japan. Laboriously, they made their way to Mason City, Iowa.

The corporate team was convinced that they could manufacture glutamate and sell it in the Orient and in the U.S. through the Larrowe-Suzuki Company. Financially the deal was disastrous for Larrowe, but he was determined to manufacture MSG from his waste water.

In 1931 Dr. Ikeda died. Subsequently, in 1936, Suzuki died and the Larrowe-Suzuki partnership ended. Larrowe, in his seventies and in poor health, had never made his waste water truly profitable despite enormous infusions of his own wealth. Dr. Albert Marshall, a good friend and consultant to James Larrowe, was most impressed by Larrowe's tenacity in finding a use for the waste water. In fact, he said of Larrowe, "Of him it can be said in the words of Shakespeare, 'For what I will, I will and there an end.'"

Profit or not, Larrowe had finally found a use for his waste water; the manufacture of monosodium glutamate. His company, the Amino Products Corporation, was eventually sold to the International Minerals and Chemicals Company.

Despite American manufacture, MSG was still not an accepted ingredient in American food. Ironically, the Japanese use of Aji-no-moto in rations was eventually noticed by the U.S. soldiers during World War II. After the war, conferences were held to discuss use of this flavoring agent particularly in prepared foods for the army and field rations, as well as for the fledgling frozen food industry.

By 1948, the first symposium about this flavoring secret was held at the Stevens Hotel in Chicago and presided over by the Chief Quartermaster of the Armed Forces. In order to understand the enthusiasm generated by this meeting we have to return to 1948.

Imagine yourself as an influential member of the food industry called to Chicago in 1948 to hear about an apparent wonder substance—an Oriental food secret which even made Japanese army rations more palatable. Entering the symposium hall you are greeted by the erudite Franklin Dove, the chief of the Food Acceptance Branch of the Quartermaster Food and Container Institute of the Armed Forces. Colonel Charles Lawrence, the commanding officer of the Quartermaster Institute, greets you with pleasure and assures you that like the others called to this symposium he has been awaiting this event with great expectations. "I have enormous curiosity about this strange substance," he says. As you look around the meeting room you see leaders of the food industry and representatives of almost all the major food processors and sellers in America. The name tags emphasize their arrival from all parts of the country, many after an arduous train ride, some after long automobile trips and a lucky few after a bumpy airplane ride.

Here on your left you recognize Mr. Bowers from Campbell Foods, on your right Dr. Anson from Continental Foods. John Andrews from General Foods is in front with a contingent of several people in tow. Behind you is Mr. D'Sinter from Standard Brands and Mr. Guest from the Continental Can Company. The few women present are Dr. Charlotte Dalphin from Nestles, Miss Inez Jobe from the United Airlines Food Service, and Anna Boller from the National Livestock and Meat Board. As you look around the packed room you see representatives from Borden, Stokeley, Libby, Pillsbury, and even Oscar Mayer.

This meeting marked an American revolution in food, an eight-hour event with consequences for millions and millions of people over many years. What was it that started this food revolution? These are some of the actual excited comments:

"MSG suppresses undesirable flavors"

"MSG gives zest to food"

"Increases flavor and odor appeal"

"Suppresses bitterness and sourness"

"Removes the tinny taste from canned foods"

"Enhances acceptability"

"Meaty, chicken-like flavor"

"Flavor appeal improved particularly with meats, seafoods, stews, and chowders"

"I wouldn't believe that the same product that could improve the taste of marinated herring would also improve lobster paste, crab cake, and codfish cakes, but it is true and it does"

"Creates a lingering flavor reaction"

As you look about the hall you see people taking notes furiously as each speaker gives glowing reports of the wonder substance. All through the hall is a buzz of excitement. MSG has made a hit. The United States MSG food industry is off and running. And if all the remarkable findings aren't enough, then what the last speaker of the day offers is the icing on a marvelous cake.

Dr. Carl Pfeiffer of the University of Illinois College of Medicine, a well-known physician who has taken the time to come to this symposium, now tells the assembled group that scientists are beginning to do experiments to test the effects of glutamate in raising the IQ level in feeble-minded people. The assembled food people scarcely need any more incentive, but this news provides a turbo boost.

America had truly discovered MSG, or to be more accurate, MSG had discovered America. Perhaps Al Capp could have been found in that room to be inspired to base his "Schmoo" on MSG. The Schmoo was a wonderful and loving animal which, when you were hungry, would taste like your favorite meal.

Had you been at this symposium you would have emerged from the meeting room into the chilly spring Chicago air with an overall enthusiasm for a very special ingredient.

The United States food industry responded with vigor. It is now difficult to find a product without added MSG. Little did these enthusiastic food producers know that their "miracle substance" had a dark side which would eventually result in the MSG Syndrome.

CHAPTER THREE

No Laughing Matter

At first many people thought it was a joke when Dr. Ho Man Kwok wrote to the *New England Journal of Medicine* in 1968 reporting his reaction to MSG. "The syndrome," he wrote, "which usually begins 15 or 20 minutes after I have eaten the first dish, lasts for two hours without any hangover effect. The most prominent symptoms are numbness at the back of the neck gradually radiating to both arms and the back, and general weakness and palpitation." Another letter writer urged the real Ho Man Kwok to come out and admit his practical joke and his real name, "Human Crock," but Dr. Kwok was real and his symptoms were very real.

In 1969, Dr. Herbert Schaumberg at the Albert Einstein School of Medicine began a careful scientific study of MSG effects. Nevertheless, in an ironic vein, he wrote, "To suppress the mounting hysteria and prevent the wholesale slaughter of Chinese restaurant owners we feel impelled to present a preliminary communication on the etiology, psychopathology and clinical pharmacology of the variously misnamed post-sino-cibal syndrome (Chinese restaurant syndrome)." With tongue in cheek, Dr. Schaumberg thanked the untold numbers of victims who had called him in the middle of the night. He even proposed to write a book titled "The Race for the Double Wonton," the proceeds to be used to "improve working conditions for the dedicated [Chinese restaurant] workers."

On July 1, 1968, Franz Ingelfinger, the editor of the *New England Journal of Medicine,* wrote, "Whatever its best name, the reaction to 'Kwok's disease' as described in our issue of May 16th has uncovered a legion of hitherto silent sufferers. Many apparently had been sitting all tingling and tormented but not saying anything, each utterly sure that he alone was wretched and never dreaming that the contented-looking co-consumer of bird's nest soup likewise was in agony. Even husband and wife wanted to spare each other."

Yet, how funny was it really? As Dr. Herbert Schaumberg and his colleagues investigated further they identified some symptoms which could have been produced by ingestion of MSG–burning, facial pressure, chest pain, and headache. They discovered that one bowl of soup (six to seven ounces) could precipitate symptoms in an MSG-sensitive individual, and this occurred at a dosage of three grams or less–a smaller amount than commonly used for flavor enhancement.

When 56 normal people ranging in age from 21 to 67 years were tested (30 male, 26 female), symptoms of the syndrome occurred in all but one subject. And this individual did develop symptoms when MSG was administered intravenously.

One of the particularly disturbing symptoms that Schaumberg's study found was pressure throughout the chest area, occasionally radiating to the arm or the neck. This alarming sensation caused one of the test subjects, a physician, to request an electrocardiogram, fearing that his symptom was that of a myocardial infarction (MI) or heart attack. (In recent medical literature, emergency medical technicians have been warned to consider differential diagnosis–"MI or MSG"–when the patient's presenting symptoms include sweating, numbness around the face and neck, chest pressure, and burning sensations, palpitations, nausea and vomiting [*EMS,* May 1986, 14(4)].)

In his 1969 paper in *Science,* the official journal of the

American Academy of Science, Dr. Schaumberg concluded, without humor, that "MSG can produce undesirable effects in the amounts used in the preparation of widely consumed foods."

Eight years later Dr. Liane Reif-Lehrer, a noted researcher at Harvard Medical School, conducted a questionnaire study of reactions to MSG. Of the 1,529 respondents, almost 30% reported reactions including dizziness, nausea, abdominal pain, visual disturbances, fatigue, shortness of breath, and weakness (*Federation Proceedings,* April 1977).

In Dr. Reif-Lehrer's study the most commonly reported symptoms were headache and tightness around the face. An appreciable number of subjects reported dizziness, diarrhea, nausea, and stomach cramps. Fifty people described eye symptoms ranging from burning eyes and blurry vision to pressure and seeing lights or colors. Many people had emotional reactions ranging from depression or insomnia to feeling "tense." The sample was predominantly of adults, but she also used 14 classes of grade school children, a total of 317 in grades two through six in two different towns. These children were asked about types of feelings after eating various foods. She concluded that 19% of children interviewed had some reaction to MSG, with nausea and stomachache being most common. While reactions of the questionnaire group generally subsided within three to four hours, ten percent reported longer lasting symptoms.

Dr. Arthur Colman, a respected psychiatrist in San Francisco, began collecting data after he detected two dramatic cases of reactions to MSG. One particularly striking case, as reported to him by the child's father, a physician, was that of Noah:

"Noah was nine years old when we noticed that he often developed headaches and stomachaches after fast-foods. He also had a difficult time holding back his urine and his stool. Sometimes he would even go in his pants. This was very difficult for him at the age of nine. He also had periods of hyperactive-like behavior, difficulties which came and went with nothing apparently causing or setting off the problem.

"His pediatrician was at a loss. In general, Noah was a charming, gentle, smart and friendly boy. But when he had these periods he became irritable, whining, yelling and unreasonable. The school noticed his changes and we met with his teacher who described how he was usually very responsive and friendly, but had been experiencing periods when he seemed to change into another child. We called it 'Devil to Angel' and were very confused. His difficulty with diarrhea and controlling his bowels had an impact on his whole personality, because he couldn't play with other children during these episodes. He began therapy with a child psychiatrist after his pediatrician couldn't find any causes for his behavior.

"We noticed that his problems became worse after eating Kentucky Fried Chicken and finally discovered that MSG was used in large quantities in the batter and the secret spices. For the next two weeks we tried to keep Noah away from any food containing MSG. His symptoms cleared miraculously and he had a dramatic improvement in his bowel control.

"One day the family ate at a Chinese restaurant and Noah had a big bowl of wonton soup. Within five minutes he developed a face flush and within a few minutes felt a terrible cramp and need to rush to the toilet. For the next week his symptoms all returned including his hyperactivity. We felt that the cause was MSG and we continued to be very careful with his food.

"We didn't tell the child psychiatrist about our discovery and Noah kept going to therapy. At our next month's conference his psychiatrist asked what we had done because Noah seemed to be a different child. The psychiatrist advised terminating therapy if the situation continued in improvement. Several weeks later he advised that we stop psychiatric treatment. At our conference with Noah's school the teachers had noticed a tremendous change in his consistency and in his personality.

"His moods now seem so much more genuine and not driven. There is no way to describe the difference in our sense of joy in the family since we took Noah away from MSG."

Were the symptoms confined to a mild reaction without lasting or serious results one could easily dismiss the effects of MSG. But a reaction like Noah's is seriously life and socially disabling. In some cases, reactions can be life-threatening as well.

Dr. Dietmar Gann from the Division of Cardiology at the Mt. Sinai Medical Center in Miami Beach, Florida, wrote in the July 1977 issue of the *Southern Medical Journal:*

"A 36-year-old man came to an emergency room complaining of severe weakness, palpitations and sweating. He had been in good health with no medical problems until he went to a restaurant. Approximately one half hour after beginning his meal with wonton soup he developed a burning and tight sensation in his chest, face and neck. He began sweating and felt weak. A friend who was dining with him also developed some weakness and tingling in the arms, but the symptoms were much milder.

"He drove home and shortly after developed chest pain and pressure with the pain radiating to both shoulders and upper arms. The chest pressure lasted about one half hour and then subsided, but the weakness remained and the man was brought to the hospital.

"An electrocardiogram showed ventricular tachycardia and he was treated with lidocaine, a medication to reduce the excitability of the heart. Within three minutes the rate and rhythm had returned to normal. Overnight hospitalization showed no further changes and tests revealed no heart damage."

Dr. Gann's conclusion was that MSG can produce not only a feeling of chest pressure and pain, but it can cause serious heart irregularities in susceptible individuals.

Dr. David Ratner, Dr. Elyakim Eshel, and Dr. Ehud Shoshani reported on their cases in the *Israel Journal of Medical Sciences* in 1984. One 52-year-old man was admitted twice to their coronary care unit because of severe upper abdominal pain and a burning sensation in his chest. Full medical evaluation failed to show a cause. The physicians tested him with MSG and he developed shortness of breath, numbness of the legs, heat in his face and chest, and abdominal swelling and pressure. Follow-up revealed that since avoiding MSG he has been symptom-free.

Other examples of the range of MSG reactions show a wide range of problems from mild annoyances to severe debilitating symptoms, as illustrated in the following case histories.

Case Histories

Migraine-like headaches were experienced by a 40-year-old female, S.B. She relates her story:

"In the 1960s, I was having debilitating migraine headaches two or three times a week. No pain killers got rid of the migraine completely—it just had to wear off. A couple of times a doctor even came to my house and gave me an injection to put me to sleep. I was in my early twenties and living in California. For the most part the doctors were at a loss for a cause and patted me on my back saying, 'Now, dear, you have to quit letting things get to you.'

"One day I read about the 'Chinese Restaurant Syndrome.' What a revelation! For lunch I was having lunchmeat sandwiches or canned soups. Then I'd use Accent salt when preparing dinner. Avoidance of MSG brought dramatic *results.*

"I find my efforts to avoid MSG are difficult when travelling or eating at friends' houses. I once ate two sausage patties and was sick in less than an hour. The reaction usually lasts almost 12 hours: flashing lights, nausea, light-sensitive eyes, pounding pain (always in the same spot over the right ear and in my right temple), weakness and slurred speech.

"In 1980 I had surgery. My first food after the procedure was beef broth, which I refused to eat until I reviewed the ingredients. It took the hospital over three hours to produce the ingredients (which of course contained MSG) and much persuading before they would bring me a different meal. I couldn't believe that the hospital dietician hadn't heard of MSG!"

L.M., a 49-year-old female, relates the following:

"About a year ago I was having a set of symptoms that led me to see two physicians (an internist and a psychiatrist). My symptoms were wide *mood swings, headaches, and hot flashes. Being 48 at the time, I thought it might be menopausal symptoms. My doctor put me on a light concentration of estrogen. The hot flashes, headaches, and mood swings were lessened, but I still experience them periodically, and they seem to be related to food and fatigue.*

"My headaches come about 45–60 minutes after I eat things like fast-food hamburgers. Many of my headaches are at the base of my skull

where the spine connects. It's a throbbing headache. Other symptoms that I experience are stomachaches and diarrhea, muscle aches all over my body, dizziness and trouble with vision. I am 49 but feel like I am 80.

"In recent years I have suspected I might be reacting to MSG because of an increase in stomach cramps, headache, and diarrhea after eating Oriental food. I have tried to eliminate it from my diet by avoiding foods labelled with MSG, but still continue to have occasional headaches, stomachaches, and listlessness after eating some foods. I realize now that I must not only avoid MSG, but also ingredients such as hydrolyzed vegetable protein and natural flavors."***

MSG reactions can drastically affect peoples' life styles, their emotional states, and their physical well being. One of the most drastic, long-term reactions is related by R.E., a 45-year-old woman. A particularly distressing aspect of her reaction was her arthritis-like symptoms. The next case accentuates the potentially disabling effects of the joint and tendon symptoms:

"About eight years ago I became ill with a low-grade fever and muscles that hurt so bad that I had trouble moving out of a chair. We thought it was the flu and I missed an entire week of work. The fever subsided at the end of the fifth day and the muscles were less sore until I used them. Even normal activity like walking up and down stairs caused problems.

"I am a very active person. From that time on until now, every time I exercised (even singing more than normal), I would be very sore after a short rest. I could still do activities such as golf, but couldn't get out of the car when we returned home from the golf course. My eyes seemed strained after a short time of reading, my singing voice was acting like I was over-using it, and my hands would get sore just holding a knife for a short time or when playing the piano or typing on the computer. What was so strange was that the muscles didn't bother me as long as I was using them, only after I stopped.

"I was checked for lupus, multiple sclerosis, muscular dystrophy and

*Contains up to 20% MSG.
**May contain amounts of MSG.

arthritis. The doctor put me on beta blockers along with a diuretic for stress and a borderline high blood pressure.

"My present doctor suggested I see a physical therapist to deal with the symptoms. She would give me a massage to relieve the muscle spasms It helped for only a short while—until I did more than just sit or stand. I forced myself to walk approximately two miles a day just to keep up my normal level of exercise so I wouldn't hurt just to move.

"One day we went to an Oriental restaurant for dinner. I carried the leftovers home and ate them for lunch the next day. During the 24 hours that I ate the Oriental food I became increasingly sore and had a bad case of diarrhea. In desperation I called my physical therapist and she asked me what I had eaten in the last 24 hours. She suggested that I locate everything in my kitchen cabinet that contained MSG and not use it for the next five days. At the end of five days I was not sore anymore for the first time in eight years.

"Every time I eat out I can tell within half an hour if I have ingested MSG inadvertently by a return of the diarrhea. Some restaurants are not very good at telling you if they use MSG. If I get a dose it takes exactly five days for it to work out of my system.

"I discovered that all my regular seasonings, soup mixes, sauces, salad dressings, meats, etc., were laced with MSG. Needless to say, I don't use them anymore and I am free of the constant pain and muscle locks. We have a hot tub which I used for stress and over-use of muscles, but the hot tub is not needed anymore for me to be able to function normally. Paring vegetables and fruits causes no problems either. Golf and cross country skiing are again joys. Dancing and singing are again pleasures.

"It is a pleasure to be free again and know what is causing a problem when it happens (which is rare now). I have learned what foods contain MSG and ask very pointedly about the contents. Eating, of course, is not a carefree sport anymore, but I feel good."

There is a syndrome involving muscle and joint aches which can be mistaken for arthritis. The next case of a 55-year-old female pediatrician is very similar to that of R.E.:

E.B. had tendonitis and arthritis of all her joints, particularly in her lower extremities. For a period of 15 years she underwent extensive test-

ing, but no cause could be determined. At times it was necessary for her to use a walker to move about her home.

One evening she went with her family to an Oriental restaurant for dinner. The following day she could not move about, even with the aid of the walker, due to extreme swelling and pain.

At this time she suddenly considered MSG as a possible cause. Tracing her symptoms to MSG in Oriental food and also common foods gave her "a new life."

It took a large dose and a 15-year history to uncover the cause of her suffering. E.B. remarked that "because of this symptom I lost so much time and lost out on interviews. My life now might have been very different if I had known [about MSG]."

Also of some interest is that her daughter, a 24-year-old medical student, experiences great lethargy after ingesting MSG.

While ingestion of MSG does not always cause such severe, debilitating reactions, other responses can be very unpleasant. L.P. is such an example:

L.P. is a healthy 40-year-old physicist, born in China. He takes no medications and has no known allergies. He developed episodes of a rash over his entire body, with circular areas ranging from quarter-sized circles to large circular areas over six inches in diameter. The rash would clear up in a week or two, but it was so uncomfortable that he consulted a dermatologist who diagnosed the condition as "atopic dermatitis."

With scientific precision, L.P. began to keep a log of his activities to discover the common rash-causing factor. After several months it was clear that the rash occurred within a day of his eating at Oriental restaurants. After experimentation L.P. determined the cause of his rash was definitely MSG, as the rash did not reoccur when this additive was eliminated from his diet, and would return whenever he inadvertently ate MSG-containing foods.

L.P. also experienced outbreaks of rash after long flights on commercial airlines. This had been diagnosed by another physician as an "altitude rash," but meals served on-board planes frequently contain MSG (see Chapter 9). L.P. had assumed, as many do, that MSG is found only in Oriental cuisine.

Another typical case was reported by D.J., a 39-year-old male:

"I'm divorced, living alone in a small apartment and have been frequenting a new Chinese restaurant. I started out slow in my consumption, but have gradually been increasing my visits to three or four days a week. I began to notice in the afternoon after eating there that something wasn't so pleasant.

"I would become depressed and my stomach would be churning and a very distinct aftertaste would develop. I've also been having some sharp headaches. They are always centered just above my right temple."

E.S., a 49-year-old woman, works as a physical therapist. Her reaction illustrates how some of the severe symptoms can appear like a stroke:

She first noticed some reaction to MSG after eating soup at a Tahitian restaurant when she was in her early twenties. At that time she found her speech to be slurred, began to shake and felt as though she would lose consciousness. Approximately ten years later E.S. was eating an Oriental soup on an empty stomach. She experienced these same symptoms along with a tremendous feeling of weakness and depression which took several days to resolve.

She believes that as she has aged her tolerance to MSG has decreased and that she is now much more sensitive. E.S. notices bloating and weakness at much smaller doses than were previously tolerable. When she ingests a large dose her speech is slurred and her balance affected. The reaction always takes several days to clear.

Many of the case reports received contain such vague symptoms as swelling, fatigue and moodiness, tension across the forehead, ache in the jaws and temples, and headaches. One elderly woman reacts with severe hives and difficulty waking after ingesting MSG. Others report diarrhea and vomiting or sneezing and nasal stuffiness.

As stated in a summary by A.S., a 38-year-old man, the reaction to MSG can be momentarily severe and frightening:

"In 1973 in Boston I was out for lunch with three fellow employees

and our boss. We chose a Chinese restaurant for our meeting place. As we were leaving the restaurant the boss's face became flushed. He complained of chest pain and felt he was having a heart attack.

"All four of us got excited and decided to rush him to a hospital. On the way I noticed that my jaw was particularly locked up. I couldn't open my jaw all the way, a symptom I had heard could be related to MSG overdose.

"The delay of time due to heavy Boston traffic allowed us to see that the boss's condition was not as serious as we had thought.

"Since that time, I have experienced this locking jaw a few more times after eating at a Chinese restaurant."

Monosodium Glutamate Intolerance in Children

While it has become clear through clinical studies that at least some adults react adversely to monosodium glutamate at levels commonly found in processed foods, and almost all will react as the dose is increased, children were not at first used as subjects for test experiments. However, 19% of the children interviewed by Dr. Reif-Lehrer reported adverse symptoms and ten percent of the adults who reported symptoms remembered having them also during childhood.

As a precautionary measure the manufacturers of baby food removed MSG in 1969 when early tests with experimental animals cast some suspicion on its safety. Yet, while MSG is no longer found in baby food, it is a common element in the diet of children who eat ordinary canned and processed meals, not to mention fast-foods and school lunches. There is no regulation whatsoever on the use of MSG in school cafeterias despite its possible ill effects on some children. Dr. Liane Reif-Lehrer became particularly concerned in 1976 when she encountered three children whose serious symptoms were traced to this food additive.

The first case involved a child who developed "shudder"

attacks at the age of six months when he was started on adult foods. The attacks continued and the child was thought to have a form of epilepsy. Yet, the medications ordinarily used for seizures were ineffective in his case. These symptoms of "shudder-shiver" were finally traced to MSG in the child's diet. The symptoms soon stopped after the diet was changed. They returned during a period of trial feedings with food known to contain high amounts of MSG.

Her next case, reported in the *New England Journal of Medicine,* involved a 16-month-old child who developed what were described as "shivers." The symptoms disappeared when her diet was changed to eliminate monosodium glutamate.

Dr. Reif-Lehrer also described a 14-year-old boy who had experienced episodes of intense headaches and vomiting since he was ten. He had been diagnosed as having "a migraine-like syndrome or seizure equivalent." Eventually his symptoms were traced to monosodium glutamate and the episodes were controlled.

One grateful mother wrote to Dr. Reif-Lehrer:

"Because of your studies on monosodium glutamate, my son Todd h been relieved of severe migraine-like headaches and stomachaches. Todd started this behavior at about age ten and was finally hospitalized at a 11 for all kinds of tests. These showed nothing but a slight electro-encephalographic abnormality, not uncommon in children. The neurolo gist sent him home with a short list of foods to eliminate. Monosodium glutamate was on the list.

"Todd has had only one slight case of this behavior since that day and this was because he ate some chicken noodle soup at scout camp. W watch carefully for this ingredient in any foods and avoid it.

"Thank you so much for spreading the word. . . ."

Many other letters of this sort were received. Yet had the word really been spread? The letter about Todd was written in 1978. Since that time the amount of MSG added to foods has been increased.

Asthma and MSG

Asthma may be precipitated by MSG. Drs. Allen and Baker from Australia published a report in the *New England Journal of Medicine* about two young women who developed severe asthma episodes after ingestion of MSG at Chinese restaurants.* Originally these doctors were skeptical about MSG being the precipitating factor. They gave capsules containing 2.5 grams of MSG to the young women (a bowl of wonton soup contains approximately 3 grams). In both cases there were asthmatic reactions. The first case responded to simple treatment. The other young woman had severe unresponsive asthma, requiring that she be placed on a ventilator.

Not content with their initial experiments because they were carried out on only two people, the scientists decided to expand their study. Working from the department of Thoracic Medicine in the Royal North Shore Hospital in Sydney, Australia, Dr. David Allen, Dr. John Delohery and Dr. Gary Baker worked for three years to establish the link between MSG and asthma.

Their results were reported in the Journal of Allergy and Clinical Immunology *in October, 1987, and are alarming.* Over a three-year period of time they tested 32 patients with asthma. These patients were selected for testing because of either a history of an attack after a Chinese restaurant meal, or "unstable" asthma with sudden severe unexplained attacks. Also, some patients with a past history of being sensitive to other chemicals were chosen (for example, prior reaction to sulfites or aspirin). The asthma reactions were not confined simply to mild shortness of breath—as the following case illustrates.

A 23-year-old registered nurse was presented to the emergency room at 8 a.m., 11–12 hours after ingestion of a 15-course Oriental meal. On awakening that morning her asthma progressed in severity and did not improve with her usual inhaled medications designed to dilate the airways (called broncho-

*See Appendix 2 (p. 109) for the "first person" account of this important discovery.

dilators). Because of her worsening condition, full intravenous treatment was initiated in the hospital, including giving adrenaline, cortisone and broncho-dilators to her, but they did not help. Close to death, she required intubation (a tube inserted into her windpipe), artificial ventilation (by a machine ventilator) and partial cardiopulmonary by-pass. Finally, after five hours of this intensive treatment, she began to improve.

When later tested with a dose of MSG, she developed an asthma attack. She was tested with less than one-half teaspoon of MSG and developed asthma and stomach pain after 12 hours. Naturally she was very cautious, as were her doctors, and they agreed that only one test dose be administered and the dose be such that she might get it from food. It was critical to determine if MSG was producing her asthmatic attack. In the evening she was given two and one-half grams of MSG (less than one-half teaspoon). She did not have asthma through the night and the following morning she was well until around 10 a.m. when a sudden severe asthmatic attack occurred. Within one half-hour she again required intubation and ventilator support for five hours.

Dr. Allen and his colleagues placed her on a strict diet to avoid MSG and to report on that regimen. "This patient's asthma is now well controlled. She has stopped oral cortisone, rarely requires admission to the hospital and has not been ventilated in two years." While this patient's case was particularly dramatic, Dr. Allen and his associates present 13 cases of asthma which they concluded were induced by MSG. All in all, of the 32 patients studied, 16 appeared sensitive to MSG.

Using a "challenge dose" of pure MSG, the scientists then performed tests to measure lung function. As a test dose, they chose one-fourth to one-half teaspoon (one and a half to two and a half grams)—a dose which would commonly be taken from ordinary food. Their results add to the serious questions about MSG and asthma. Of the 32 patients, 13 responded to the test doses of MSG with marked reductions in their air flow. An asthma attack developed in seven people within one to two hours and, in the other six people, attacks occurred within 12 hours.

This careful landmark study leaves little room for doubt. The conclusion that MSG can provoke an asthmatic attack is inescapable. The scientists were also able to show that the reaction is worse as the dose increases and an attack may be delayed as long as 12 hours.

Many asthma sufferers are seen in emergency rooms. How many of these cases are precipitated by MSG has never been analyzed. Rarely are patients with asthma told to avoid MSG or change their diet.

According to government statistics, the overall death rate from asthma increased 23% between 1980 and 1985. The rise "astonishes medical professionals," according to *The Wall Street Journal* (Wednesday, February 18, 1987, p. 29) and the cause of this rise is unknown and disturbing. Ten million Americans suffer from asthma and the National Institute of Allergy and Infectious Diseases terms it "the leading cause of absenteeism among schoolchildren." The incidence in children aged three to 17 is rising rapidly. Some of the increase is related to MSG—but just how much is not known.

Depression and MSG

In 1978 Dr. Arthur Colman identified a depressive reaction in a 38-year-old woman which was traced to monosodium glutamate. His patient's reaction involved a prolonged psychological effect lasting two weeks.

Elisabeth was a 38-year-old professional woman and mother of two children who was in good health unless she ate food containing MSG. Approximately half an hour after she ate this additive she noticed a feeling of tightness around her neck, pressure behind her eyes, a mild headache, some flushing, and mild nausea and stomach pains. She went to her physician because of her symptoms and an extensive evaluation showed no ulcer or intestinal disorder. These attacks came frequently and, at first, no cause could be established.

Eventually a more troubling symptom developed. She began to have a short-term depressive reaction. Elisabeth would feel gloomy and then would believe that other people were speaking to her with sarcasm and unpleasantness. Her whole family noticed these episodes with her typical facial expression and downcast spirits. Within a week or two this strange reaction would disappear.

Finally her husband began to trace her reactions to the food she was eating and discovered the cause to be MSG. When she succeeded in eliminating MSG from her diet, all symptoms stopped. On some occasions the symptoms recurred and in each instance MSG was found in one of the foods she had consumed.

Discovery of the cause of Elisabeth's symptoms made a dramatic difference in her life. As her husband said, "It would be a mild statement to say that life has been better for Elisabeth and our family since the cause of her distress was discovered. The depressive episodes are entirely gone and the emotions she experiences feel to us and to her more natural. It is far easier to become genuinely angry or sad when one does not feel in the grip of a toxic substance. The continual abdominal and stomach distress has entirely gone."

Dr. Colman concluded, "There are obvious neurochemical clinical and social implications of this psychiatric syndrome."

After the case report was published, Dr. Colman received dozens of letters from other people suffering symptoms which were traced to MSG.

Depressive symptoms may vary, but all depressive conditions share certain features: sadness, slowed motor activity, less self-confidence, and what the late Dr. Nathan Kline, a leader in the evaluation and treatment of depression, considered the cardinal symptom—anhedonia (absence of pleasure). Put simply, anhedonic people fail to experience the usual satisfaction from their life and activities. The striking case report of Elisabeth aroused enormous interest because there are an estimated 100 million people worldwide who suffer from clinically recognized depression. Teenage and childhood depression are also on the rise, with increased suicide rates.

The rising incidence of depression in younger age groups has resulted in the coinage of a new phrase, "Agent Blue," as reported in *The Atlantic Monthly* in December 1986. "More people are becoming depressed and at much younger ages. The ailment is so prevalent that some National Institute of Mental Health scientists privately speculate that an unknown 'Agent Blue' may be spurring its spread." With the tendency of children and teens to eat processed foods at home and at fast-food restaurants, their dose of MSG is increasing. Symptoms induced by this flavor enhancer may be a factor in the rising rate of depressive conditions.

One of the greatest dilemmas encountered by physicians is the proper treatment of depression. Some advocate drug therapy and others counselling or psychoanalysis. Still others recommend shock therapy.

How many of the estimated 100 million sufferers are experiencing the depressive symptoms reported by Dr. Colman as caused by MSG? And in how many of these cases is MSG identified as the real culprit? Very few of the physicians and psychologists who treat depressed people are aware of this possible cause. Yet research leads to the conclusion that many people are experiencing mood changes directly associated with MSG.

Dr. Reif-Lehrer's questionnaire study indicated that 30% of her subjects were aware of experiencing some reaction to MSG. This finding has been confirmed by many other investigators—that amounts commonly added to many foods induce MSG-sensitive symptoms. My own studies conducted through interviews with patients and questions tabulated from lecture audiences have confirmed this figure. Between 25–30% of the people I have questioned believe they have had reactions to MSG. In addition, when the question is asked, "Do you react to MSG or know someone who does?", the positive response is almost unanimous. In Chapter 5 there is a symptom checklist which can be used by readers to determine if they belong to this group.

Depression is a subtle psychological symptom in its earliest form, but chemical theories support the possibility of MSG involvement. Some scientists have speculated that the "Chinese Restaurant Syndrome" is caused by a release of acetylcholine in the body. Experiments have shown that acetylcholine may induce depressive reactions. MSG has also been found to be toxic to brain cells in animals.

However, regardless of whether the biochemical mechanism can be pinpointed at this stage, sensitive people can react to MSG with a depressive response. Sometimes it is short-lived, a matter of hours. In the initial case report of Elisabeth, the 38-year-old woman's depression lasted as long as two weeks and caused marked life derangements—almost to the point of divorce.

When mood changes occur, people will frequently act upon their mood swings. If they become sad and irritable, those around them suffer. One sensitive reactor I interviewed noticed how he became depressed and angry. "Until I discovered my reactions to MSG I was always getting into arguments after eating at Chinese and fried chicken restaurants. We once went to marital counselling [several days] after such an argument. But by the time we got there . . . the reaction had cleared and I couldn't even remember what made me so annoyed."

Depression is the most common psychiatric disorder treated by physicians in office practice. The late Dr. Nathan Kline stated that "more human suffering has resulted from depression than from any other single disease affecting mankind." Since depression has been seen for hundreds of years it is obvious that it has causes other than MSG. Yet, from our research and case studies, the conclusion is inevitable that some of the increasing depression in our society is, in fact, caused by increased ingestion of MSG.

Because of these concerns some physicians have begun to uncover and identify cases of MSG reactivity, particularly those with behavioral change and depression as an important component.

Many cases, such as that of Dr. A.R., a California physician, have now been reported:

A 44-year-old physician and allergist, Dr. A.R., developed a sensation of dizziness and whirling. This symptom continued until it became disabling and the doctor had to be driven to work every day. He also noticed a feeling of great weakness about an hour after lunch.

Eventually Dr. A.R. consulted a neurologist. His symptoms at that time were, in addition to a whirling sensation, difficulty in reading, depression, diarrhea, and frequent headaches. The neurologist sent Dr. A.R. to the hospital for a complete neurological work-up including brain scans and electroencephalograms. These proved to be negative.

Dr. A.R. finally traced his symptoms to a seasoning salt he had been using that contained monosodium glutamate. He also was taking in MSG from many other sources. The symptoms disappeared about a week to ten days after elimination of MSG from his diet. In fact, Dr. A.R. reported that after eliminating MSG from his diet he felt better than he had in years. Whenever he accidentally consumes the flavor enhancer the symptoms reappear.

Dr. A.R. warns, "I have seen many patients with chronic headaches, depression and fatigue. These patients are most grateful when they find that their symptoms go away when they avoid MSG."

Why do such symptoms occur? In Chapter 6 we will delve further into the human biochemistry and some of the animal experiments which point to the brain sites of the reactions. However, for the subtle behavioral changes in humans, we must study people.

Dr. Reif-Lehrer was an associate professor in the Department of Ophthalmology at Harvard Medical School when she first became interested in glutamate effects. She is what is usually termed a "basic scientist"—that is, a scientist who concentrates upon research into basic effects in isolated tissue cultures under carefully controlled conditions. Having seen MSG toxicity on an experimental level (after adding it to chick retinas and tissue cultures), she began to wonder if monosodium glutamate was affecting children. After interviewing 300 children she concluded that about 20% reported reactions which appeared to be MSG-related.

Because her questionnaire study of 1,500 adults found that approximately 30% had adverse reactions to foods containing added MSG, she was called to testify on glutamate before the Federation of American Societies for Experimental Biology Select Committee on GRAS (Generally Recognized as Safe) substances. She concluded in her testimony in July 1977 that "because of the results of some of the available laboratory studies and because there is a sizeable amount of information indicating that an appreciable number of people do perhaps react adversely to elevated levels of glutamate, it would seem the better part of wisdom not to have unrestricted use of this material until further research has been done concerning possible subtle and long-range effects of this amino acid in humans."

While acknowledging that "there is evidence that some individuals may respond to relatively small doses" and recognizing that "there should be some constraint placed on the addition of MSG to processed foods," the Select Committee made no further findings. The principal reason given for their not taking a more vigorous anti-MSG stand was that it did not cause brain damage in experimental animals. But how do we measure depression or mood change in such an animal? What about headache, flushing, chest tightening, and the myriad symptoms reported by human patients that are difficult or impossible to identify in non-speaking subjects?

The report further notes that MSG is listed on food labels, and so people sensitive to it can avoid it. Right? Wrong!

Hiding MSG

What could be better for us than protein? And everyone knows vegetables are good for us. Thus, "hydrolyzed vegetable protein" sounds safe and even wholesome. However, this is the chemical method of producing monosodium glutamate.

A mixture of hydrolyzed proteins contains the salts of other proteins as well, and monosodium glutamate may comprise as much as 20% of hydrolyzed vegetable protein (the usual range is 12–20%). The flavor enhancement produced by this mixture is almost entirely dependent on MSG.

Few people are aware that products containing hydrolyzed vegetable protein frequently are advertised as "all natural." While MSG must be specifically listed on food labels, hydrolyzed vegetable protein, which contains MSG, may be designated simply as "natural flavorings."

Headaches and MSG

Headache seems to be a common symptom of MSG sensitivity and a major study is now under way at the respected Diamond Headache Clinic in Chicago to further detail this association. While most of the headaches reported have been mild, the following case of a young woman shows that MSG-related headaches may be socially disabling:

Jackie is a 21-year-old graduate student who had experienced occasional headaches for many years. She had received full neurological tests including EEG and brain scan, but these showed no apparent cause for her complaint. In other respects her health was good.

When Jackie became a graduate student she began to spend more time in study, at the library and at home. She also began to eat more frequently at fast-food restaurants. At this time the severity and frequency of her headaches increased and she needed narcotic type medications to control them. She began using Percodan almost on a daily basis.

Eventually her physician suggested that diet might be a factor in her headaches and he suggested she avoid Chinese food. The headaches began to lessen in frequency after a few days, but still they persisted.

At that point Jackie began some investigating on her own. She discovered that the fast-food hamburger chain she frequented added MSG to their hamburger mix and that it was also added to the canned tuna fish she liked and the sauces at her favorite restaurant. Even the onion soup at an exclusive French restaurant in her locality was laced with "protein powder." She discovered that packages labelled "No MSG or preservatives" contained such ingredients as "Kombu-extract," i.e., MSG, and "vegetable powder," i.e., MSG in hydrolyzed vegetable protein. In short, these labels were misleading.

Motivated by the continued severity of her headaches, Jackie methodically removed all MSG from her diet and began eating a diet of fresh foods at home. Within two weeks her headache problem of several years had almost completely disappeared.

She still gets occasional headaches when she is upset or under tension, but she says that these are very different in severity and duration from those she formerly experienced. She reports that her life has been improved in many areas, and her social life in particular.

Other Symptoms Associated with MSG

Aging

Many people who are sensitive to the effects of MSG have reported that they could eat the additive with impunity when younger, but noticed increasing reactivity with age. With decreasing ability to shop for fresh foods, particularly in winter, older people often must rely more on processed, canned, dried, and pre-packaged food. With the current level of MSG addition to such foods (either as MSG or as hydrolyzed vegetable protein), this ensures a steady MSG diet which is a factor in edema and depression. (The frequency of depression in the elderly has often been noted, but the association between such states and diet is rarely made.)

Other symptoms associated with MSG sensitivity include dizziness and balance difficulties, common problems of the elderly. In larger concentrations, MSG is a potent nerve toxin and it has been theorized that chronic long-term ingestion may be involved with Alzheimer's disease and Parkinson's disease, as well as other nerve cell degenerative diseases, such as amyotrophic lateral sclerosis (Lou Gehrig's disease).

A summary of the experimental evidence linking glutamate and human disease was compiled by Dr. J. Timothy Greenamyre at the University of Michigan Neuroscience Laboratory and published in *The Archives of Neurology* in October 1986. Dr. Greenamyre noted that an injection of a glutamate-like substance can produce many of the abnormalities of Huntington's disease (i.e., bizarre movements and mental deterioration), and suggests that glutamate neurotoxicity may be involved in the cause of this disease.

Dr. Greenamyre also proposed that some of the brain cell degeneration found in Alzheimer's disease (a form of senility and dementia) may be caused by increased stimulation of parts of the brain by glutamate. An elevated glutamate level has been associated with one type of brain cell atrophy (olivo-ponto cerebellar atrophy), causing severe balance difficulties.

With these associations it may be wise for people with a family history of nerve cell degenerative disease to limit their intake of MSG as early as possible.

MSG and Women's Cycles

Some women of child-bearing age who are sensitive to MSG also associate this substance with worsening of pre-menstrual syndrome symptoms. PMS affects many women in various degrees and generally includes swelling, mood changes, depression, irritability, and general malaise. The reported cases clearly suggest that individual testing is necessary to show which cases of PMS are worsened or possibly even initiated by the use of monosodium glutamate.

CHAPTER FOUR

The MSG Syndrome: What You Don't Know Can Hurt You

The evidence points to MSG sensitivity of epidemic proportions among the general population. But for most people the symptoms are vague: depression, headache, mild nausea, pressure around the eyes, tingling of the face, behavioral disturbances in children, and mood swings, accentuating those already present in adolescents and also in adults. In the previous chapter, some striking cases were described, and the studies of Reif-Lehrer, Schaumberg, and others clearly have shown that we are dealing with many sufferers of the MSG syndrome—literally an epidemic.

The usual causes of an increase in symptoms are changes in the environment or in human behavior. Recently we have witnessed important changes in personal habits. Families eat together less often. When they do sit down to a family meal, it often consists of convenience foods such as fried chicken, frozen dinners, canned soups, vegetables in sauces and prepared salad mixes, all of which usually contain MSG. When you further consider that monosodium glutamate is an ingredient of seasoning salts, bouillon, hydrolyzed vegetable protein, meat tenderizers, most prepared spaghetti sauces, most sausages, some bacons, and even such ethnic foods as gefilte fish and matzo balls, it becomes quite evident that the level of MSG in any family meal may be quite high. In fact, the giant food company

Pillsbury just completed a year-long study focusing on American eating behaviors from 1971 to 1986. The results of their study appeared in the *Wall Street Journal* on Tuesday, March 15, 1988 (p. 39). The largest and fastest-growing segment of the U.S. population is what Pillsbury calls the "Chase and Grabbits." They are most likely to subsist on fast-food, frozen dinners and carry-out pizza. Betsy Morris, the staff reporter, concludes, "This group which swelled 136% over the 15 years of the study now comprises 26% of the population."

According to a study by NPD Group, a Port Washington, New York, marketing firm, only 15% of meals in 1986 required the use of a standard oven. This means that pre-packaged foods are much more prevalent. Linda Smithson, director of Pillsbury Company's consumer center, said, "The fragmentation of eating has become much more normal than random."

MSG is a standard ingredient in most canned soups, soy sauces, bouillon cubes, soup stocks, and frozen dinners. It is difficult to avoid even in the finest of restaurants. The amount of MSG consumed daily by the average person has risen to a level where more and more people are experiencing harmful reactions. According to the latest Encyclopedia Brittanica (1985), even the cigarette smoker may get additional MSG since tobacco leaves have been cured with MSG for smell and flavor enhancement.

What is the basis for the popularity of MSG? It is technically termed a "flavor enhancer" and that is what it does. But it can also mask inferior freshness and quality in the foods we eat.

The "tinny" taste of canned foods is diminished and its color maintained with the use of MSG. So the advantages of its use for the food industry are obvious, and for those people who do not react to the substance except in very large doses its addition is innocuous. Yet for the millions of people who experience adverse symptoms at the present usage level the problem is serious.

In most cases we are not dealing with an allergic reaction. It is instead a classic toxic reaction. The difference requires some explanation.

An allergic reaction, experienced only by some individuals, is usually not dose-dependent and involves the body reacting to a substance in a particular way, causing the body to release certain chemicals. This can result in asthma, skin rashes or itching, or even obstruction of respiration due to swelling of structures within the neck and the airway.

On the other hand, everybody will react to a toxic substance, and the degree of reaction depends on the dose. Take, as an example, caffeine. Caffeine in coffee will affect everybody who drinks coffee. The only variables are the tolerance level of each person and the amount of caffeine in the coffee bean. One person who is very sensitive will say that one-fourth or one-half of a cup of coffee will make him or her feel jittery or sweaty, and perhaps cause some heart palpitations. Others feel the effects of caffeine when they take a whole cup; still more people will react to two cups and more at four. As the number of cups of coffee taken over an hour or two increases we will reach a point where almost every human being will experience at least some effects of the caffeine.

The situation is similar with MSG. The intensity of the symptoms produced by MSG and the number of people reacting will increase as the dose is increased.* Even when asthma is precipitated by MSG, the reaction is dose-related (that is, the higher the dose, the worse the asthma attack). Thirty percent of the population will react when given five grams of MSG and 90% will react when given ten grams. The particular symptoms produced depend on the individual, but there are characteristic patterns. In one day of eating processed and fast food, an individual can easily consume five grams or more, exceeding his or her symptom-causing dose.

The quantities discussed may appear small, since there are approximately 30 grams to an ounce. One-fifth or one-sixth of an ounce of MSG may not look like much, but it has potent effects.

*There are those who are particularly sensitive and will react to less than one gram of the substance. An average bowl of wonton soup contains three to five grams, one teaspoonful of Accent salt contains almost six grams.

In addition, consumption of MSG has doubled in every decade since the 1940s. More than 80 million pounds are used yearly in the United States alone, and this doesn't even include the enormous amounts of hydrolyzed vegetable protein added to foods.

Despite the widespread nature of this syndrome, the cause is often hard to trace. Even when the cause is discovered it can take years and even decades for people to act on it. For example, scurvy began to increase with the growth of sailing and shipping. From the fifteenth century to the nineteenth century more than two million deaths were attributed to scurvy. James Lind, a Scottish physician, demonstrated the cause in the late 1700s as lack of fresh fruit. Yet, 50 years after his "proof" ships routinely left port for prolonged sea voyages without a supply of fruit.

Modern medicine accepts the idea of deficiency diseases such as scurvy, but it may be overlooking the effects of toxic substances such as MSG. We now have a substantial body of largely ignored evidence that MSG may cause troublesome reactions for many people.

It has been almost 20 years since the first case report was published in the *New England Journal of Medicine.* It is more than 18 years since Dr. Herbert Schaumberg, Dr. Robert Byck, Dr. Robert Gerstl, and Dr. Jan Mashman reported in *Science* that "MSG can produce undesirable effects in the amounts used in the preparation of widely consumed foods." It is more than ten years since extensive questionnaires showed MSG sensitivity to be widespread. Yet many people are unaware of the cause of their MSG-induced symptoms and more MSG is being used in foods than ever before.

As the use of MSG has become more widespread, case reports are coming from all parts of the world: from France, Germany, Australia, Israel. Even in Japan where MSG has been a staple of cooking almost since its discovery in 1908 people have been aware of reactions to this substance. The MSG syndrome is spreading.

CHAPTER FIVE

MSG Symptom Analysis

Are You Reacting to MSG?

There are three general categories of symptoms resulting from MSG. These are allergic, peripheral, and central (brain).

The allergic symptoms include characteristic allergic responses such as asthma, skin rash, or sneezing. Peripheral responses are flushing, tingling, chest tightness, palpitations, or headaches, and arthritis-like symptoms of the joints. The central symptoms arise from the brain and include depression, mental confusion, insomnia, or restlessness.

Table 1 categorizes common symptoms, Table 2 shows children's symptoms, and Tables 3, 4, and 5 attempt to divide the symptoms according to type. However, one person can have a mixture of the different types of symptoms of the MSG syndrome.

TABLE 1

Symptom Analysis

Gastrointestinal
Cramps
Diarrhea
Nausea

TABLE 1 (continued)

General and Chest	*Other*
Tightness around face	Rash, hives
Tingling/burning in face and chest	Chills, shakes
Tightness in chest, chest pain	Tenseness
Headaches	Numbness of face
Weakness	Confusion
Dizziness	Water retention
	Speech slurred
Eye Symptoms	Paranoia
Blurring of vision	Sneezing
Seeing shining lights	Thirst
Difficulty focusing	Muscle aches
Tingling around eyes	Jaw stiffness
	Sleepiness
	Asthma
	Staggering
	Depression
	Heaviness of arms and legs
	Stiffness
	Excessive perspiration
	Fast heartbeat
	Insomnia
	Balance problems
	Arthritis
	Tendonitis

TABLE 2

Children's Symptoms

Behavioral problems	Chest discomfort
Stomach cramps	Thirst

TABLE 2 (continued)

Headache	Nausea
Stomachache	Dizziness
Tiredness, depression	Throat symptoms
Loss of bowel or bladder control	

TABLE 3

Allergic Symptoms

Rash
Hives
Asthma, shortness of breath
Sneezing
Running nose

TABLE 4

Peripheral Type Symptoms

Flushing
Jaw tightness
Headache
Rapid heartbeat
Chest tightening
Diarrhea, stomach cramps
Arthritis

TABLE 5

Central Type Symptoms (Brain)

Depression
Insomnia
Confusion
Paranoia

Foods to which MSG has been added have more effect when taken on an empty stomach. The Chinese Restaurant Syndrome, better named the MSG Syndrome, often is manifest immediately after the first soup course. The presence of carbohydrates and bread in the stomach tends to modify and diminish the reactions.

Mechanisms of MSG Toxicity: Why is it Toxic?

Glutamic acid is one of the amino acids which make up proteins. In nature, the glutamic acid is linked by "peptide" linkages to other amino acids. When a person eats a protein substance, the linkages are broken apart slowly in the digestive process. When pure MSG is given, a rapid effect occurs from the glutamate. This "free glutamate" is not attached to other amino acids. The normally slow breakdown process is bypassed because there are no "peptide" linkages to slow the process. The sudden increase in glutamic acid within the body is rapidly absorbed and can raise the normal blood level of glutamate to eight or ten times its usual amount.

CHAPTER 6

MSG: Bad Taste

Taste relies primarily on the perception of stimuli through the taste buds located in small depressions on the tongue. When stimulated, these taste buds transmit electrical signals to the brain which interprets the variety of signals and forms an overall impression. The brain is like a television set receiving a variety of signals. Eventually an image is formed which is perceived as "taste." MSG placed on the tongue stimulates these electrical discharges, making the "picture" more intense.

MSG has become popular because it enhances every kind of flavor and increases smell appeal. It has been found to increase saltiness and accentuate sweetness, and reduce sourness and bitterness. It modifies undesirable tastes, such as the "earthiness" and raw peel flavor or the "fishy" taste of lima beans. In addition, MSG has been found to intensify "bloom"—the spreading of taste through the mouth—and the aftertaste that stimulates the next bite.

A flavor enhancer is different from a seasoning in that it can intensify existing flavors without adding a strong flavor of its own. Although MSG does have a flavor of its own (chicken-like, sweet-salty, sometimes perceived as bitter), it is a relatively weak and subtle flavor. This aspect makes it very difficult for an MSG-sensitive person to know when he or she has inadvertently ingested MSG or hydrolyzed vegetable protein.

So just how does MSG work? It appears to increase the sensitivity of the taste buds and feeling receptors in the mouth, and stimulates an electrical discharge. Basically, it is a mouth aphrodisiac.

In addition, it has actions on intestinal muscle, blood vessels and in the brain.

Glutamate as a Neurotransmitter

Some of the increased recent concern about glutamate stems from the observation that it acts as an excitatory neurotransmitter in the brain. A neurotransmitter is a substance which stimulates brain cell activity.

In 1957 in Baltimore, Dr. D. Newhouse and Dr. J. P. Lucas used glutamic acid in an effort to reduce hereditary retinal degeneration in rats. At that time, they were proceeding on the idea that glutamic acid might have a protective effect. Their findings were just the opposite—that MSG resulted in rapid irreversible destruction of the majority of cells in the retina within minutes. After this initial observation was scientifically validated, other scientists, particularly Dr. John Olney of Washington University in St. Louis, determined that glutamate could produce not only retinal damage, but damage to other parts of the brain. The earliest evidence of damage is to the "dendrites" which are the filaments which transfer electrical information. Damage to these is widespread. Subsequently, the nerve cells are altered and die. Dr. Olney focused his work on the damage to the hypothalamus which regulates overall body and glandular function.

Animal Studies of MSG: Implications for Humans

Animal studies are useful in determining the effects of substances on living organisms, but generally such studies use much higher doses, proportionately, than those usually taken by people. Consequently, they do not relate to human symptoms unless the higher dosages lead to observable brain damage.

Dr. John Olney's research showed that the MSG brain injuries in rodents resulted in obesity, behavioral disturbances, endocrine hormone changes, stunted bodies, seizures, and infertility.

How these powerful effects on experimental animals might relate to possible effects on humans is not known. Dr. Olney and other scientists considered that caution in the addition of MSG, particularly to foods consumed by children, is advisable.

Excesses of glutamate have been linked to a condition called Huntington's Chorea which involves destruction of parts of the brain. Symptoms mimicking those of amyotrophic lateral sclerosis (Lou Gehrig's disease), such as muscle weakness and slurred speech, have been produced by large doses. Some scientists have speculated that the rising incidence of Alzheimer's disease may be related to MSG. Because of the known brain cell neurotoxicity of MSG, they theorize that chronic long-term exposure might lead to neuronal (brain cell) loss. This certainly is an area for further scientific study.

Although there may be long-term effects, these are still just theorized. What is of most importance at this time are the myriad symptoms in humans caused by this substance—some of which can be disabling or life-threatening.

Scientists studying these problems have sometimes run into severe obstacles. As Dr. Baker, the Australian asthma specialist, notes, "The story needs to be told. This will include not only the side effects of MSG, but also the discrediting of the people honestly reporting these effects, the frustration many of us have had in finding funds to do the necessary further research and the anger at the regulatory bodies for ignoring this problem."

CHAPTER SEVEN

Living Without MSG: MSG-Free Cooking at Home

For the MSG-sensitive consumer an obvious question arises: how to prepare nutritious meals, within a reasonable amount of time, and still avoid MSG?

The increasing availability of natural foods provides one answer. But many consumers are intimidated by natural food stores. Not only are many of their products more expensive, but the ingredients used may be unfamiliar.

There are a growing number of convenient, pre-packaged natural foods available. However, not all products sold in health food stores, or all those labelled "natural ingredients," are free from MSG or hydrolyzed vegetable protein. Also, in several Oriental products the kombu seaweed or extract used for flavoring is high in natural MSG (see Chapter 8, p. 57).

The pre-packaged and pre-frozen foods which generally contain substantial amounts of MSG are undeniably convenient. But cooking without MSG doesn't have to be a time-consuming life vocation. You can prove this for yourself by trying out these time-saving techniques that include easy substitutes for your favorite recipes that may contain products with MSG, and numerous make-ahead recipes that may be used in small portions over a period of time (see Appendix 1).

Avoiding MSG may take a little more time and attention, but your own well being and that of your family surely is worth a little effort. As for those who are very sensitive to the substance, an MSG-free diet can change your life.

Easy Food Shopping Tips

1. Think fresh! The more fresh ingredients you can incorporate into your meals, the more likely you are to avoid products containing MSG. Check into local farmers' markets or cooperative food markets. These can be a good source for fresh foods at inexpensive prices.

2. Read labels carefully (see Chapter 8 for tips on reading labels). If the ingredients include hydrolyzed vegetable protein, natural flavoring, flavorings, vegetable protein or vegetable, chicken or beef broth, proceed with caution. If you are unsure if a listed ingredient in one particular product includes MSG, check other products of the same manufacturer. If kidney beans from that manufacturer lists "spices," while their barbequed version of beans lists both "spices" and "MSG," the "spices" in their kidney beans should be safe.

3. Before purchasing any delicatessen or smoked meat products check the label. Most sausages and some luncheon meats will contain some form of MSG. Do not hesitate to ask your butcher or delicatessen clerk to let you read the labels.

4. Avoid processed or dried foods with "flavor packets." These packets almost always contain MSG to improve the flavor. They are most commonly found in boxed rice and pasta dishes, powdered salad dressing mixes, and dried soups. Also keep in mind that food which has been dried has lost a good deal of its nutritional value.

5. Canned gravies, chili, stews, and sauces frequently contain large amounts of MSG and other additives. Read these labels carefully.

6. Check the ingredients in your favorite canned fish or meat products.

Substitutions

Using MSG-free products in your favorite recipes can be an interesting challenge. In the recipe section in the appendix you will find many of the old-fashioned basics that were used for generations before the processing and frozen food revolution.

These easy recipes will substitute for many of the prepared items we have come to depend upon, such as canned or condensed soups.

Canned Mushroom Soup: Use the same amount of the Quick White Sauce with mushrooms.

Catsup: Homemade catsup (see recipe) or the same amount of tomato sauce. If using tomato sauce you may want to add some seasonings, since prepared catsup is generally spicier than plain tomato sauce.

Canned Chicken Stock or Bouillon Cubes: Several stock recipes are included in the appendix. The herb stock is much faster, but the others have a richer flavor. Plain water can also be substituted, especially in soup recipes. Spices and herbs will round out the flavor.

Barbeque Sauce or Chili Sauce: Either homemade catsup or tomato sauce with added spices or chili will provide a substitute. Recipes are also included for homemade barbeque sauce or chili sauce.

Monosodium Glutamate: If a recipe calls for the addition of MSG there are a few alternatives. Fresh lemon juice works well on salads, soups, tomato products, sauces, vegetables, meat, and fish dishes. Herbs and spices are a wonderful addition to most foods and will easily replace MSG.

Soy Sauce and Worcestershire Sauce: Tamari (available in many health food stores) is lower in salt and MSG than soy sauce, but is not completely MSG-free. Look carefully at the ingredients in worcestershire sauces—they vary greatly from one brand to the next. When stir-frying vegetables or meats a liquid of ginger, lemon juice and water can be substituted for the soy sauce liquid. This will give the dish a lighter and fresher taste.

If you must use canned products, favor the low-sodium variety. They may not be free from MSG, but the likelihood is that it may be used in smaller quantities.

When using spices for seasoning, avoid the pre-mixed combinations of spices, garlic salt, and onion salt. Some companies add MSG to these products.

Another labor-saver is the Food Preparation Network: a group of friends who get together and divide up a week's cooking chores. One prepares enough homemade bread for everyone for one week, another prepares two varieties of salad dressing, while yet another makes soup, and so on.

Just imagine a Monday morning when you give four couples or individual friends a large jar of homemade barbeque sauce that you spent two hours making on Sunday afternoon and you receive five loaves of whole grain bread, one gallon of soup, two quarts of salad dressing, and half a gallon of tomato sauce along with a jar of catsup. Not bad for two hours' work and a lot of fun among friends.

Some of the recipes found in the appendix are for large quantities. These can be cut in half depending on your needs. Remember, however, that it is convenient to have soup or bread ready in the freezer for those cold winter nights.

CHAPTER EIGHT

Shopping Smart and Reading Labels

As a general rule, the more processing a food undergoes, the more likely it is to contain MSG. It is also true that the ambiguities of labelling requirements can make it difficult to determine whether or not a food product contains MSG.

Some foods are better complemented by MSG than others. Listed below are foods grouped in order of their likelihood of containing monosodium glutamate.

Most Likely

Potato chips and prepared snacks
Canned soups and dry soup mixes
Canned meats, box dinners, and prepared meals
Frozen foods (seafood, chicken, and dinner entrees)
International foods
Poultry injected with broth

Very Likely

Diet foods
Salad dressings
Cured meats and lunch meats

Less Likely

Cookies, crackers (unless seasoned), and candy
Dairy products
Frozen vegetables
Breads, pasta, baking supplies (unless MSG is added to the dough)

Seldom or Never

Ice cream
Frozen juices
Soft drinks and citrus juices (except possibly V-8)
Canned fruits and vegetables
Cereals, pancake mixes, and syrups
Bulk foods (rice, grains)
Fresh meat and fish

MSG and Food Labelling

The Food and Drug Administration (FDA) regulates the labelling requirements of food products in the United States. Packaged food products are required by law to have their ingredients printed on the package, with the ingredients listed in descending order of predominance.

If MSG is added to a food product during its processing it will be listed on the label. However, it is permitted to list "hydrolyzed vegetable protein" on the label without mentioning that it may contain up to 20% monosodium glutamate. Hydrolyzed vegetable protein or protein hydrolysate (as it is sometimes called) is classified by the Food and Drug Administration as a natural flavoring. Therefore, a packaged food item might specify "natural flavoring" in its list of ingredients without mentioning that the natural flavoring consisted of hydrolyzed vegetable protein—which contains MSG.

Monosodium glutamate is on the GRAS (Generally Recognized as Safe) list of the Food and Drug Administration. It is not considered a food additive, but an unregulated seasoning classed by the FDA as a flavor enhancer.

Section 101.22 of the Federal Code of Regulations deals with labelling requirements for spices, flavorings, colorings, and chemical preservatives. Part H(5) states: "Any monosodium glutamate used as an ingredient in food shall be declared by its common or usual name monosodium glutamate." Therefore, if monosodium glutamate is used as an ingredient in the processing of a food it must be labelled as such. However, hydrolyzed vegetable protein need not have its MSG content listed.

The Federal Code of Regulations Section 101.22, defining natural flavorings, states that "natural flavoring" be applied to: "the essential oil, oleoresin, essence or extractive, *protein hydrolysate,* distillate or any product of roasting, heating or enzymolysis which contains the flavoring constituents derived from a spice, fruit or fruit juice, vegetable or vegetable juice, edible yeast, herb, bark, root, leaf or similar plant material, meat, seafood, poultry, eggs, dairy products or fermentation products thereof whose significant function in food is flavoring rather than nutritional" (italics added). This definition paves the way for labelling protein hydrolysate as "natural flavoring," a designation that will mislead the average consumer trying to avoid MSG.

The following is a list of names to check for on labels.

Monosodium Glutamate Aliases

Accent
Ajinomoto
Zest
Vetsin
Gourmet powder
Subu
Chinese seasoning
Glutavene

Glutacyl
RL-50
Hydrolyzed vegetable protein (12–20% MSG)
Hydrolyzed plant protein
Natural flavorings (can be HVP)
Flavorings
Kombu extract
Mei-jing
Wei-jing

Figures 1–17 provide illustrative labels from some representative foods.

Many foods come packaged without additives but with an accompanying flavor pack in which the additives are contained. For MSG-sensitive people, the product may be used and the flavor pack discarded.

This flavor pack, for example, contains monosodium glutamate and hydrolyzed vegetable protein (which contains MSG). Two other ingredients here, disodium inosinate and disodium guanylate, are potent flavor enhancers which may act in a similar fashion to MSG, although there have been no large-scale studies of their effects.

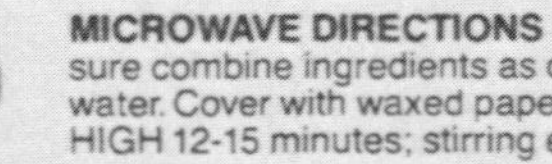

Many sauce mixes contain MSG–"Homemade" or not.

Even in health food stores, labelling can lead to false conclusions. Here, the product label states "No MSG or Preservatives." Yet under the ingredients section we see "kombu extract," which contains MSG, and "vegetable powder," which most likely refers to hydrolyzed vegetable protein (also containing MSG).

This spaghetti and meat balls product is a lunch-time favorite. While MSG is listed only once on the label, it is also contained in the "hydrolyzed plant protein" and possibly in the "flavorings." This can potentially provide a triple dose for children.

Most people would not suspect that tuna fish is packed in a solution containing MSG. This label indicates the "hydrolyzed protein" clearly. MSG-sensitive people should avoid all tuna fish packed in solutions containing hydrolyzed protein. Some are packed in a "vegetable broth" which is also suspect.

Seven Seas

Buttermilk Recipe®

Buttermilk Recipe® dressing is the original pourable buttermilk dressing made with 100% fresh lowfat buttermilk, minced onions, garlic, parsley and pepper. Try dipping chicken pieces in Buttermilk Recipe® dressing, season, cover with bread crumbs and bake. It also makes a perfect vegetable dip, baked potato topping and fried shrimp batter.

INGREDIENTS: PARTIALLY HYDROGENATED SOYBEAN OIL, CULTURED LOWFAT BUTTERMILK (CULTURED LOWFAT MILK, SODIUM CITRATE), SUGAR, SALT, WATER, DRIED ONION, LACTIC ACID, MONOSODIUM GLUTAMATE (ENHANCES FLAVOR), XANTHAN GUM (IMPROVES POURABILITY), DRIED GARLIC, POLYSORBATE 60 (IMPROVES CREAMINESS), SORBIC ACID (A PRESERVATIVE), EGG YOLK SOLIDS, NATURAL FLAVORING, SPICES, BHA AND CALCIUM DISODIUM EDTA (AS PRESERVATIVES).

SHAKE WELL
Refrigerate After Opening

BEST WHEN PURCHASED BEFORE

Anderson Clayton Foods
DIV. ANDERSON, CLAYTON & CO.
DALLAS, TEXAS 75266-0037

49880.04-P

Many salad dressings contain MSG. This buttermilk dressing contains MSG even though dairy products usually do not combine well with MSG. Other possible sources of MSG in this product include the natural flavoring.

At a salad bar an MSG-sensitive person should avoid all pre-mixed dressings as well as artificial bacon bits, prepared salads, and seasoned croutons.

The Lawry Company uses MSG in many products. Hydrolyzed vegetable protein is used in many seasonings. "Garlic salt" is one product which many people believe to be garlic and salt. However, it too may contain MSG.

The frozen food industry has used MSG or MSG-containing substances almost since its inception. On this label monosodium glutamate is clearly listed among the ingredients. It is also likely to be found as a component of the "chicken broth" and in "natural flavorings" as hydrolyzed vegetable protein.

Chicken soup with rice is another favorite of children. Hydrolyzed vegetable protein and monosodium glutamate give young children a double dose of MSG—a dose that is larger proportionate to their body weight than it is for adults. Whereas MSG has been removed from baby foods, many infants and small children are being fed soups such as this one, resulting in unsuspected MSG ingestion.

Many recipes require "beef broth" or "beef consomme." The prepared broths usually contain high concentrations of MSG in the form of added MSG and added hydrolyzed vegetable protein as well as natural flavoring.

Restaurants which claim to be "MSG-free" may be using such canned beef consomme or chicken broth.

MSG is often added as a short-cut or to disguise inferior broth. The no-salt (usually MSG-free variety) of the canned product will indicate its actual quality.

This is an example of a company using MSG without saying so. They are safe in claiming to use "No Preservatives" and "No Artificial Ingredients" since MSG is not considered either a preservative or an artificial ingredient. In this product the "chicken stock" may contain MSG in some form. The "hydrolyzed vegetable protein" contains MSG and more hydrolyzed vegetable protein may be an ingredient in the "natural flavoring."

Most canned soups contain MSG in some form. In this product, the MSG is most likely contained in the "beef stock" and is certainly found in the "hydrolyzed vegetable protein." The term "natural flavoring" can be confusing since it can also refer to hydrolyzed vegetable protein. The term "spice," while non-specific, refers to the flavoring herbs such as basil, rosemary, ginger, horseradish, etc., as well as pepper.

Even kosher foods contain MSG which, being considered a salt, is cleared for orthodox use. The MSG in these frankfurters is in the hydrolyzed vegetable protein which is listed as "flavorings." MSG-sensitive people can get marked reactions from most hot dogs.

"Snack foods" frequently contain MSG. These include most flavored potato chips and crackers. When both the chip and the dip contain MSG, the snacker may be ingesting it at unsafe levels.

Many recipes call for mushroom soup and dips are often made using these soup or flavor packs. The "no preservatives, no artificial ingredients" listed on the label do not *indicate an MSG-free product. In this instance, MSG is added in its pure form and as an ingredient of the hydrolyzed vegetable protein.*

Many restaurants believe that there is no MSG in their tuna and other prepared salads. MSG can be present in the tuna fish or in the added substances in the various "Helpers." The term "natural flavor" often indicates hydrolyzed vegetable protein. Note also that whereas sodium sulfite has been banned from salad bars, it is still found in products such as Tuna Helper and may continue to find its way to salad bars in this fashion.

This is an example of a label that may conceal the MSG content of its product. The list of ingredients includes the word "flavorings" which may indicate hydrolyzed vegetable protein, employed primarily for its MSG content (generally 12–20%). Because hydrolyzed vegetable protein contains a mixture of salts of amino acids, MSG is only one component. Many food producers are not aware that HVP contains MSG.

MSG and the Supermarket

For those who are MSG-sensitive, shopping in a supermarket can be a challenging experience.

There are a wide range of food items that can contain MSG. As stated above, some of those food items have MSG marked clearly on their labels, while the MSG content of others will be more ambiguously labeled "hydrolyzed vegetable protein" and "natural flavorings."

The labels of some spices list "monopotassium glutamate" instead of monosodium glutamate. They usually are advertised as "sodium free." However, the glutamate ion they contain affects the taste buds and the nervous system. Those with adverse reactions to monosodium glutamate are therefore cautioned against these products as well.

The following list was compiled from food products sold in five representative supermarkets. All of these food products contain some form of MSG or hydrolyzed vegetable protein.

The list is by no means exhaustive. Many food products containing MSG in some form are not listed. However, this sampling should give some idea of the range of products which contain monosodium glutamate.

Canned Meats, Prepared Dinners, and Side-Dishes

Bumble Bee Canned Tuna, packed in water (HVP)*
Chicken of the Sea Canned Tuna, packed in water (HVP)
Geisha Solid White Tuna
Chicken of the Sea Canned Tuna, packed in oil (HVP)
S&W Chunky Tuna, packed in water (HVP)

Hormel Chunky Turkey
Hormel Chunky Beef
Polar Crab Meat
Swanson Chunky Chicken Spread (HVP)

*HVP indicates hydrolyzed vegetable protein.

Underwood Chunky Chicken Spread
Underwood Roast Beef Spread

Betty Crocker Oriental Classics Chow Mein (HVP)
Betty Crocker Oriental Classics Stir Fried Rice
Chun King Chicken Chow Mein
Chun King Shrimp Chow Mein
Chun King Beef Pepper Oriental

Austex Beef Stew
Chef Boyardee Beef Ravioli
Chef Boyardee Spaghetti and Meat Balls
Chef Boyardee Smurf Pasta in Spaghetti Sauce
Chef Boyardee Tic-Tac-Toes
Franco-American Beef Raviolis in Meat Sauce
Franco-American Spaghetti and Meat Balls
Hormel Chili
Kraft Tangy Italian Style Spaghetti Dinner
Lipton Egg Noodles and Sauce (Beef or Chicken)
Betty Crocker Hamburger Helper
Betty Crocker Pork Stuffing Mix
Betty Crocker Rice Oriental Hamburger Helper
Betty Crocker Mushroom Chicken Helper
Frenches Creamy Stroganoff Potatoes
Golden Grain Chicken Flavor Rice A Roni
Golden Grain Beef Flavor Rice A Roni
Kellogg's Croutettes Stuffing Mix
Stovetop Americana New England Style Stuffing Mix
Stovetop Corn-bread Stuffing Mix
Stovetop Mushroom and Onion Flavor Stuffing
Stovetop Chicken Flavor Stuffing
Swanson Chicken and Dumplings
Uncle Ben's Country Inn Rice Florentine
Uncle Ben's Broccoli Rice Au Gratin
Vigo Yellow Rice Dinner

Amore Tortellini with Cheese
Amore Raviolini

Canned Soups and Dry Soup Mixes

Baxter's Scotch Pea Soup
Bookbinder's Shrimp Bisque
Borden Soup Starter
Campbell's Chicken with Rice condensed canned soup
Campbell's Beef Noodle condensed canned soup
Campbell's French Onion condensed canned soup
Campbell's Home Cookin soups
Campbell's Quality Chicken Noodle dried soup mix (HVP)
Campbell's Quality Onion Mushroom dried soup mix (HVP)
Campbell's Quality Onion dried soup mix (HVP)
Crosse and Blackwell Onion Soup
Knorr Chicken Flavor Noodle flavored soup mix
Knorr Country Barley Soup
Knorr French Onion flavored soup mix
Knorr Hearty Vegetable flavored soup mix
Lipton Chicken Noodle dried soup mix
Lipton Beef Vegetable dried soup mix
Lipton Onion dried soup mix (HVP)
Maggi Bouillon Cubes
Pepperidge Farm Black Bean Soup
Pepperidge Farm Lobster Bisque
Pepperidge Farm Clam Chowder
Pepperidge Farm French Onion Soup
Progresso Chicken Noodle soup
Sid and Roxies Conch Chowder

Lipton Cup O Soup Cream of Chicken
Lipton Cup O Soup Tomato
Sanwa Foods Chicken Flavor Ramen soup
Sanwa Foods Shrimp Flavor Ramen soup

Barth's Chicken Bouillon (HVP)
Borvil Concentrated Beef Flavor Liquid Bouillon
Wyler's Beef Bouillon Cubes
Wyler's Chicken Bouillon Cubes

Cookies and Crackers

Nabisco Cheese Tidbits (HVP)
Nabisco Chicken in a Bisket flavored crackers
Keebler Cheeblers Spicy Nachos
Keebler Tato Skins Baked Potato
Keebler Tato Skins Cheese 'n Bacon
Keebler Toasteds Bacon
Keebler Tato Skins Barbeque
Keebler Cheeblers Zesty Cheddar
Keebler Tato Skins Sour Cream 'n Chives

Cured Meats

Jimmy Dean Sausage Links
Jimmy Dean Pork Sausage
Land O' Frost Thin Sliced Beef
Land O' Frost Thin Sliced Chicken
Owens Country Style Sausage
Safeway Bacon
Swift Original Brown N' Serve Pork Links
Swift Original Brown N' Serve Beef Links
Webers Whole Hog Sausage

Delicatessen

Bueno Beef Tacos
Wilson's Ham and Cheese Loaf
Wilson's Cooked Beef Roast
Carando Hot Cooked Capicola
Best Kosher Cured Beef
Breakfast Sausage
Sinai Kosher Polish Sausage
Sinai Kosher Smoked Sausage
Sinai Kosher Italian Sausage
Sinai Kosher Kishke
Sinai Kosher Chili with Beans

Sinai Kosher Corned Beef Hash
Manischewitz Fishlets
Manischewitz Sweet Whitefish and Pike
Manischewitz Gefilte
Mother's Old Fashioned Gefilte Fish

Diet Foods

Armour Chicken Burgundy Classic Lite Dinner
Stouffer's Lean Cuisine Chicken Cacciatore
Stouffer's Lean Cuisine Salisbury Steak
Stouffer's Lean Cuisine Glazed Chicken with Vegetables and Rice
Weight Watchers Lemon Butter Sauce
Weight Watchers Thousand Island Dressing Mix
Weight Watchers French Style Dressing Mix
Weight Watchers Brown Gravy Mix with Mushrooms
Weight Watchers Chopped Beef Steak and Green Pepper in Mushroom Sauce
Weight Watchers Lasagna with Meat
Weight Watchers Imperial Chicken
Weight Watchers Veal Patty Parmigiana

Freeze Dried Foods

Back Packers Pantry No Cook Stew with Beef
Back Packers Pantry No Cook Rice with BBQ Chicken
Back Packers Pantry Dumpling with Chicken
Mountain House Potatoes and Beef with Onion
Mountain House Fisherman's Seafood Chowder
Mountain House Spaghetti with Meat and Sauce

Frozen Foods

Green Giant Broccoli, Cauliflower and Carrots in Cheese Flavored Sauce
Green Giant Japanese Style Vegetables

Ore Ida Tater Tots
Skaggs Alpha Beta Tasty Taters

Banquet Fried Chicken
Brilliant Instant Frozen Cooked Shrimp (HVP)
Fisher Boy Fish Sticks (HVP)
Fisher Boy Round Breaded Shrimp (HVP)
Gorton's Potato Crisp
Gorton's Crunchy Fish Fillets
Mrs. Paul's Light Entreé Tuna Pasta Casserole
Mrs. Paul's Light Fillets Farm Raised Catfish
Pilgrims Pride Chicken Drumsters
Van De Kamps Breaded Fish Sticks
Weaver's Chicken Sticks (crispy)

Honey Suckle White Basted Young Turkey
Hudson's Basted Young Turkey (HVP)

Morning Star Cholesterol Free Breakfast Patties

Jeno's Crisp and Tasty Sausage Pizza
Red Barron's Sausage and Pepperoni Pizza
Patio Beef Enchiladas
Van De Kamps Cheese Enchilada Ranchero

Armour Chicken Fricassee Dinner
Armour Seafood Natural Herbs Classic Dinner
Armour Seafood Newburg
Armour Dinner Classics Yankee Pot Roast
Armour Sliced Beef with Broccoli Dinner
Banquet Turkey Pot Pie (HVP)
Banquet Buffet Supper Gravy and Sliced Turkey
Banquet Chicken Pie
Banquet Cook in Bag Chicken a la King Entree
Banquet International Favorites Veal Parmigiana
Benihana Oriental Style Shrimp with Rice
Chun King Meat and Shrimp Egg Rolls
Green Giant Lasagna

The Budget Gourmet Yankee Pot Roast
The Budget Gourmet Pepper Steak with Rice
Le Menu Beef Sirloin Tips Dinner
Le Menu Breast of Chicken Parmigiana Dinner
Le Menu Yankee Pot Roast Dinner
Morton Turkey Dinner
Morton Boil in Bag Entree Beef Patty
Morton Creamed Chipped Beef
Morton Veal Parmigiana Dinner
Morton Turkey Pot Pie
Night Hawk Steak and Beans
Suzi Wan Teriyaki Beef
Swanson Fried Chicken Breast Dinner
Swanson Loin of Pork Dinner
Swanson Hungry Man Turkey Dinner
Swanson Macaroni and Beef Dinner
Swanson Turkey Dinner
Tyson Chicken Francais
Tyson Chicken Cannelloni

Gravy and Seasoning Mixes

Mahatma Authentic Spanish Seasoning
Vigo Italian Style Bread Crumbs
Schilling Sloppy Joe Seasoning Mix
Schilling Chicken Gravy dry packaged sauce
Schilling Mushroom Gravy (HVP) dry packaged sauce
Schilling Brown Gravy dry packaged sauce
Sunbird Sukiyaki Oriental Seasoning mix
Sunbird Fried Rice (HVP) Oriental Seasoning mix
Sunbird Chop Suey (HVP) Oriental Seasoning mix
Food Club Onion Gravy mix
Food Club Brown Gravy mix
McCormick Brown Gravy mix
Fishland Seafood Breading mix
Knorr All Purpose Seasoning
Cavenders All Purpose Greek Seasoning

Potato Chips and Prepared Snacks

Borden's Cheez Balls
BonTon Cheese Balls
Doritos with Nacho Cheese Flavor
Fritos Bar-B-Q Flavored Corn Chips (HVP)
Planters Potato Crunchies
Pringles Sour Cream Potato Chips
Pringles Butter and Herbs Potato Chips
Tostitos with Sharp Nacho Cheese Flavor
Wise (Borden) Puffed Cheese Doodles

Frito Lay's French Onion Dip
Frito Lay's Nacho Cheese Dip
Frito Lay's Cheddar and Bacon Dip
Kraft Guacamole Dip
Kraft French Onion Dip

Slim Jim Snacks
Lowrey's Beef Jerky
Lowrey's Pepperoni Sticks

Salads, Salad Dressings, and Croutons

Kraft Ranchers Choice Dressing
Hidden Valley Ranch Salad Dressing
Kraft "Mushroom" Salad Dressing
Kraft "Buttermilk" Salad Dressing
Good Seasons "Italian" Dry Salad Dressing Mix
Schilling "Salad Toppings"
Betty Crocker Suddenly Salad
Brownberry Cheddar Cheese Croutons
Brownberry Seasoned Croutons
Devonsheer Onion and Garlic Croutons
Lipton Cool Side Salads
Lipton Creamy Buttermilk Pasta Salad
Orval Kent Chicken Salad
Orval Kent Tuna Salad

Sauces

FanciFood Soy Sauce
Heinz Worcestershire Sauce (HVP)
Kikkoman Steak Sauce
Progresso White Clam Spaghetti Sauce
Progresso Red Clam Spaghetti Sauce
Ferrara Alfredo Sauce
Ferrara Clam Sauce

Spices and Seasonings

Accent
Lawry's Seasoning Salt
Lawry's Seasoned Lite Salt
Lawry's Bacon Onion Seasoning
Lawry's Garlic Salt
Schilling Season All
Schilling Savor Salt
Spike Seasoning (HVP)
Goya All Purpose Seasoning

CHAPTER NINE

Restaurant Guide to Food Containing MSG

Most meals served in restaurants across America contain monosodium glutamate. Both the hamburger you eat for lunch at a fast food counter and the expensive veal dish you have for dinner at an exclusive restaurant may contain MSG.

For those who react adversely to MSG, eating in restaurants can be a risky matter. An unwise choice can lead to hours, or even weeks, of uncomfortable symptoms.

Meats, vegetables, seafood, seasonings, and soups served in any restaurant may contain MSG. Col. Paul Logan of the National Restaurant Association quotes an association survey indicating that "three out of four of all good restaurant operators are now using Glutamate" (*Glutamate Manufacturers' Technical Committee Publication: The Value of Glutamate in Processed Foods*).

Inquiries by my staff to hundreds of franchised restaurants, cafeterias, and individual restaurants about their use of MSG brought a variety of responses. Many were vague in their answers, citing "trade secrets," while others listed foods they served that contained MSG but did not know it was an element in hydrolyzed vegetable protein. Some who answered my questions were genuinely concerned about the widespread adverse reactions to MSG. However, frequently the chefs did not know which sauces contained MSG or HVP (hydrolyzed vegetable protein).

Fast-Food Restaurants

The food items listed below are only a sample from selected restaurants and are representative of where MSG is found in restaurant foods. Many large chain restaurants will provide booklets upon request detailing the contents of their foods. Remember: if contents include MSG or HVP, MSG is present. Beware of foods with "natural flavorings" (NF) since they may contain HVP.

Wendy's

Crispy Chicken Nuggets
Croutons
Thousand Island Salad Dressing (HVP)
Reduced Calorie Bacon/Tomato Salad Dressing
French Style Salad Dressing (HVP)
Golden Italian Salad Dressing (NF)
Chicken Fried Steak
Fish Filet
Sausage Gravy
Ketchup (NF)
Chili (NF)

Jack in the Box

Breaded Chicken Patty
Breaded Chicken Breast Strips
Breaded Onion Rings
Buttermilk House Dressing
Garlic Roll

Nacho Topping
Pork Sausage
Secret Sauce (HVP)
Sirloin Steak
Taco Salad Meat
Tartar Sauce (HVP)
Whitefish and Crab Blend
Breaded Shrimp (NF)
Hamburger Seasoning (NF)
Ketchup (NF)
Pizza Pocket (NF)

Burger King

Chicken Specialty
Chicken Tenders
Sausage
House Salad Dressing
Whaler Fish Filet (NF)
Ketchup (NF)
Hash Browns (NF)
Reduced Calorie Italian Salad Dressing (NF)

McDonald's

Big Mac Sauce (HVP)
Barbecue Sauce (NF)
Filet for Filet-O-Fish (NF)
Ketchup (NF)
Mayonnaise (NF)

Arby's

Sharp Cheddar Cheese Sauce
Breaded Chicken Breast Filet
Breaded Fish Filet
Fresh Pork Sausage

Long John Silver's

Information refused

Taco Bell

Information not available

Pizza Hut

Information not available

Denny's

Croutons
Roast Beef
Sausage Patties
Breaded Chicken Pieces
Breaded Chicken Strips
Battered Cod
Breaded Shrimp
Breaded Onion Rings
Green Beans in butter-style sauce
Italian-style Vegetables in butter-style sauce
Ranch Dressing
Brown Gravy
Chicken Gravy
Country Gravy
Au Jus
Cacciatore Sauce
Clam Chowder
Split Pea Soup
Chicken Noodle Soup
Cream of Potato Soup
Teriyaki Glaze
Stuffing

Dairy Queen/Brazier

Seasoning Salt (used on most prepared foods and hamburgers)
Chicken Pieces
Chicken Patties

Kentucky Fried Chicken

Chicken Breading Mix
Gravies
All food products

In addition to this list, a simple set of guidelines can be used to help avoid MSG whenever dining out.

1. Request that your food be prepared without any salt or seasoning salts. Often these are added at the discretion of a particular restaurant manager or owner.
2. Avoid foods with gravies or sauces.
3. Avoid those foods that are breaded and deep fried.
4. Request that French fries be salted only with table salt—not with a pre-mixed seasoning salt.
5. Choose an oil and vinegar salad dressing over pre-mixed dressings.

Restaurants and Cafeterias

Adverse reactions to MSG first came to the attention of the American public in 1969 when the "Chinese Restaurant Syndrome" was reported. Patrons of Chinese restaurants noticed tingling feelings and tightness in their chest after eating. Their reactions were traced to the MSG in the food they were eating.

Due to the connection the public made between Chinese food and MSG, many Chinese restaurants began to advertise their food as "MSG-free." Today many Chinese restaurants in cities such as Los Angeles and New York have signs in their establishment windows proclaiming their food free of monosodium glutamate. While they may avoid adding pure MSG to their dishes, most often it is still contained in ingredients used or in the soy sauce.

Both publicly and privately owned cafeterias serving large numbers of people are likely to use MSG. Not only does it improve the taste of food prepared in large quantities, it is economically wise since it prevents large trays of food from tasting old or stale. Millions of such cafeteria meals are served to children in elementary and high schools each day. H. H. Pullium of the USDA states that "there are no federal regulations or policies which prohibit or limit the use of monosodium glutamate (MSG) in the National School Lunch program or any of the U.S. Department of Agriculture's (USDA) federally funded programs." Cafeterias serving students in universities and colleges are also likely to serve meals containing MSG, although they generally provide a wider range of choice of food, including some fresh fruits and salads.

One restaurant owner we interviewed stated that MSG was indispensable in preparing certain dishes. According to him, the time and skill required to achieve the same taste that a dash of MSG immediately produces would make the cost of that meal prohibitive.

Many restaurants that use MSG claim that they do not. They are not being dishonest—for the most part they just do not know what prepared products or additives contain MSG or hydrolyzed vegetable protein. Because of this, all restaurants must be suspect of using MSG.

There are restaurants that use monosodium glutamate knowingly as a spice. Meals are dosed with it liberally because it "perks up" the flavor of food.

Then there are those who are unaware that they use it. They do not know that it is in their brand of seasoning salt or soy sauce. MSG has been integrated into so many convenience foods used by large and small kitchens alike that it is very difficult to cook commercially without encountering some ingredient to which MSG already has been added. The same distributors that supply the supermarkets with small containers of food for home use also supply the restaurants. The ingredients are not changed because of quantity. Therefore, the chicken broth base which you avoid in the grocery store may be innocently added to a dish in your favorite restaurant.

The only way to avoid MSG when dining out is to question the chef or person who serves you. Information about where foods are prepared–whether in the restaurant kitchen or received pre-packaged–will eliminate many of the MSG-containing foods. For example, shrimp breaded in a commercial breading mixture has a high probability of containing MSG or HVP. However, if the chef takes time to mix his or her own breading mixture, it is possible that it will not contain MSG.

If the person who serves you does not know where or how the restaurant's food is made, request this information from the chef or restaurant owner. If you are still unconvinced about the true ingredients of a dish it would be wise to choose another. It can be preferable to call the restaurant first and inquire about the ingredients they use.

Airplanes and Trains

The prepared meals offered to passengers on airlines or trains are likely to contain MSG unless the individual in charge is very knowledgeable.

My interview with the chief dietician of Amtrak showed that he is aware and cautious about the use of MSG. The same cannot be said of other corporations, such as the Marriott Catering Services, which provides a good deal of airline food.

To avoid MSG when travelling (especially by air), it is wise to order low-salt meals and check the ingredients in salad dressings carefully. Seafood dinners may not be free of MSG due to the sauces or boiling solutions used in preparation. The peanut snacks on airlines could be another offender—be sure to check the ingredients. Remember that the cold cuts provided in the snack sandwiches may also contain MSG.

CHAPTER TEN

Conclusion

This is a time in which families rely on processed, pre-cooked food, often frozen or in cans. The freezing of food dates back to prehistoric man, and even now many families in Alaska utilize winter temperatures to keep their meat frozen. To combat heat, which leads to the growth of food spoiling bacteria as well as natural breakdown reactions, men have used running streams, cold caves, and even ice storage. This desire to preserve one's food in the best state led to the death of at least one epicure. History shows that Sir Francis Bacon was out of doors stuffing snow into his chickens to preserve them the day before he came down with fatal pneumonia.

In America the development of a practical refrigerator in the 1930s led directly to the frozen food industry. The United States had two million refrigerators in use by 1937 and by 1953 90% of American homes had refrigerators. Now it is very difficult to find any home without a refrigerator. In Europe the movement was slower with less than 10% of French and English homes having a refrigerator in 1954, but they, as well as Japan, have since caught up.

As the frozen food industry developed it encountered certain problems. Prolonged freezing altered the taste of foods and reduced consumer acceptance. In 1947 the industry held its first symposium on MSG and heard enthusiastic reports of its flavor enhancing effects. Soon this substance was widely adopted by food manufacturers. Now it is difficult to find a frozen prepared food to which MSG has not been added in some form.

While the frozen food industry was flourishing, the canning of food continued. The canning industry had started in the early 1800s after a French storekeeper had developed a way to extract air from jars and heat the food inside the sealed jars. Nicolas Alpert won a medal from Napoleon for this work. Shortly after, an Englishman obtained a patent for Alpert's process, but he used tins instead of glass jars.

In 1850 America began its canning industry with fish products. By 1858 Gail Borden, an enterprising dairyman, was canning condensed milk. During the Civil War, armies were fed from canned foods and by 1870 more than 30 million cans of varied foodstuffs were sold. During World War I the can opener had become a staple of most kitchens in America and by World War II the numbers of cans sold yearly was in the multi-billions.

Once MSG was "discovered" in America it was widely used by canned food manufacturers to reduce the "tinny" taste of their products, especially canned soup. The dried food industry favored MSG for its flavor enhancing properties. Nobody could have anticipated that as MSG was added to more and more canned and frozen foods eventually adverse effects from this substance would begin to develop.

MSG as a Drug

A drug is a substance which, when taken into the body, causes changes in the physiology and functioning of tissues or organs. We often use the terms "drug" and "medicine" interchangeably, although, in a stricter sense, a medicine is a "drug" which is used for some therapeutic purpose, to treat some disease or malfunction. Drugs do not necessarily have a therapeutic function.

MSG is a drug which acts directly on the taste buds, altering their sensitivity. It acts peripherally on blood vessels and the lower esophageal sphincter and it acts also on the brain and central nervous system. It is properly termed a "drug" since it has no apparent therapeutic purpose.

In fact, in sufficient doses, MSG has a deleterious effect. Because it alters a sensory modality, the taste buds, it modifies reality and produces or intensifies sensations. MSG may be used to mask unpleasant flavors, disguise spoilage, and provide a sense of freshness to food which is old. By fooling the taste buds, it can be responsible for our eating food which may produce harmful upsets.

Studies indicate that monosodium glutamate can cause marked reactions in those sensitive to the substance and that it may be at the root of a widespread increase in depressive reactions and other symptoms. These are serious drug side-effects.

Our society has become one in which drugs are used to enhance human activities and the functioning of human organs beyond their natural limits. For example, the widespread use of anabolic steroids by weight-lifters and other competitive sports participants to enhance muscular strength is well known. Similarly, cocaine is used to artificially heighten mental alertness and athletic performance. For decades amphetamines and also caffeine have been used by students and night workers to heighten mental performance.

Against this background, it becomes easier to understand how the widespread use of MSG is so carelessly and casually accepted. The ill effects of MSG may not be so dramatic as those of the "hard" drugs, but ill effects there are. Why is it that they are so little known?

For one thing, there is no question that the corporate health of many companies depends on the use of MSG. Also, for other reasons less clear, we do not see taste enhancement in quite the same light as we see the enhancement of other functions by drugs. Because of the fast-paced world and the need to prepare "quickie" meals in microwave ovens we have accepted a trade-off. This trade-off could ordinarily be rationalized as just another accommodation to "life in the fast lane" were it not for the widespread reactions to this substance. Millions of children and adults are experiencing adverse reactions to MSG and do not know the cause. It is likely that some of the increase in childhood and adolescent depression results from this added substance, and part of the increase in asthma may also be attributed to MSG.

MSG-sensitive people must recognize their sensitivity and reactivity to this flavor enhancer and then learn to live without it. It will dramatically improve their lives.

As a physician, I want to educate, not scare. In my view this is a serious health problem affecting enormous numbers of people—and the public needs to know.

ACKNOWLEDGMENTS

I would like to acknowledge the following people whose help was essential in allowing me to research and accurately present to the public this important health topic.

Charles Brunn acted not only as an excellent researcher, but also offered advice, insight, and needed encouragement. His help was invaluable.

Dr. Liane Reif-Lehrer, Boston, Mass., provided important research and studies, and offered comments on the manuscript.

Manjeet Singh, Center for Food Safety and Applied Nutrition of the Food and Drug Administration, Maryland, helped to clarify some of the FDA issues.

Mihoko Miki at the UCLA University Library, Los Angeles, California, helped with translations from the Japanese.

The Richard C. Rudolph Oriental Library at UCLA, Los Angeles, California, aided in finding information about the discovery of MSG.

Anne E. Daniels, Home Economist, PET Incorporated, St. Louis, Missouri, provided practical information.

John Tyson, M.D., Albuquerque, New Mexico, offered his appraisal from a public health and pediatrics viewpoint.

Libby Lee Colman, Ph.D., San Francisco, California, provided additional case information and manuscript suggestions.

Richard Kenney, Ph.D., Professor of Physiology, George Washington University, Washington, D.C., offered dialogue concerning difficulties he encountered in researching this area.

H. M. Kwok, M.D., Maryland, for his original contributions to the topic.

Richard J. Wurtman, M.D., Massachusetts Institute of Technology, Boston, Mass.

Russell T. Spears, M.D., Long Beach, California, for providing case reports.

Richard E. Cristol, Executive Director, The Glutamate Association, Atlanta, Georgia, for research assistance.

Rie Takashima and Hisayuki Ishimatsu, East Asiatic Library, University of California, Berkeley, helped with research and translations.

Michael Jacobson, Ph.D., Elaine Blume, and the staff at the Center for Science in the Public Interest, Washington, D.C., provided suggestions, case reports, and pertinent articles.

Nancy Hollander, Attorney at Law, Albuquerque, New Mexico, for friendship, consultation, and case finding.

The Aji-no-moto Company, Tokyo, Japan, for research help.

David Shonenberg, U.S. Department of Agriculture, Washington, D.C., aided in clarifying USDA and FDA policy.

Senator Bill Richardson (D., New Mexico), Washington, D.C., helped in obtaining information.

Gertrude Gabuten, Office of Consumer Affairs, Food and Drug Administration, U.S. Department of Health and Human Services, Maryland.

Mary Ann Wilcox, SAGA Corporation, Menlo Park, California.

Robert Gunther, Reporter for *The Press,* Atlantic City, New Jersey, for help with determining the nature of school lunches.

Claire Regan, M.S., Registered Dietician for the National Restaurant Association.

Col. Paul Logan, The National Restaurant Association, for written comments.

H. H. Pullium, Regional Director, Special Nutrition Program, USDA, Southwest Region, Dallas, Texas.

Cindy Volper, Santa Fe, New Mexico, for her recipes and help with the chapter on MSG-free cooking.

The Shoko Restaurant, Santa Fe, New Mexico, for Japanese translations.

Gerald Deighton, FDA, Rockville, Maryland.

Dr. Andrew Ebert, International Glutamate Technical Committee, Washington, D.C., for his detailed letters.

Marilyn Mabery, Santa Fe, New Mexico, for help with research and case reports.

Judith Hughes, R.N., Los Alamos Scientific Laboratories, Los Alamos, New Mexico, for help with research, case reports, and case-finding.

Patricia Murphy, Librarian, Santa Fe Public Library, Santa Fe, New Mexico.

Judith DuCharme, Librarian at the University of New Mexico Library (Albuquerque). Also, the New Mexico State University Library in Las Cruces, New Mexico.

George W. Irving, Jr., Ph.D., Chairman, Select Committee on GRAS Substances, Federation of American Societies for Experimental Biology, Maryland.

L. Mullen, Epidemiologist, Communicable Disease Center, Atlanta, Georgia, for help in determining methods of detection of MSG reaction in large populations.

Neil Hollander, Ph.D., Paris, France, for help in gauging overall MSG use in Europe and for general encouragement.

Members of the Association for the Study of MSG and Food Additives who offered information, support, and encouragement.

Mary Wachs, Editor, Museum of New Mexico Press, for help with overall book, content, and structure as well as continuing inspiration.

Particular thanks to the scores of researchers over the past several decades whose research into the nature and effects of

MSG were vital to our understanding of how the MSG Syndrome has occurred and to those MSG-sensitive people who shared their adverse reactions with us.

And, finally, I wish to acknowledge the invaluable assistance of my wife, Kathleen, who assisted and advised at every step of the research and writing and with great patience and personal sacrifice saw this project through to completion.

APPENDIX 1

Recipes

By Chef Cindy Volper

Beef Gravy

(Makes about 2 cups)

4 tablespoons beef drippings (fat) or oil
3 tablespoons flour
2 cups pan juices and/or beef stock
2 teaspoons dried oregano or thyme
½ cup thinly sliced mushrooms (optional)
1 tablespoon butter
salt and pepper to taste

1. In a saucepan heat the beef drippings or oil until hot. Gradually add the flour, using a wire whisk to avoid lumps.
2. Turn down to low heat and continue to whisk the mixture slowly until the color changes from white to deep brown.
3. Add the pan juices and/or beef stock slowly as you continue to whisk.
4. Optional: Sauté the mushrooms in butter in a separate pan and add them to the stock mixture when tender.
5. Add the dried herbs and season with salt and pepper. Simmer over a low flame until you reach desired consistency.

Poultry Gravy
(Makes about 2 cups)

¼ cup chicken drippings (fat) or oil
¼ cup flour
2 cups pan juices and/or chicken stock
¼ cup cooked giblets, diced small
½ cup thinly sliced mushrooms (optional)
1 tablespoon butter
salt and pepper to taste

1. Heat the drippings or the oil in a saucepan until hot. Gradually add the flour, using a wire whisk to avoid lumps.
2. Cook over medium heat for 4–5 minutes. Be sure to continue whisking slowly and do not allow the flour to turn brown.
3. Add the pan juices and/or chicken stock slowly as you continue to whisk. Stir in the cooked giblets.
4. Optional: Sauté the mushrooms in butter in a separate pan and add them to the stock mixture when tender.
5. Season with salt and pepper to taste. Simmer over a low flame until the desired consistency is reached.

Brown Gravy
(Makes 6 cups)

Melt in a heavy saucepan:
½ cup beef drippings
Add:
One cup Mirepoix, a blend of 1 diced carrot, onion, celery heart, combined with ½ cup crushed bay leaf, 1 sprig thyme, and 1 tablespoon minced MSG-free raw ham or bacon. The

mixture should be simmered in 1 tablespoon of butter until the vegetables are soft. When this begins to brown add:
½ cup flour
and stir until the flour is a golden brown. Then add:
10 black peppercorns
2 cups drained, peeled tomatoes or tomato puree (no salt)
½ cup coarsely chopped parsley
Stir and mix well, then add:
8 cups beef stock
Simmer for about 2 hours or until reduced by ½. Stir occasionally and skim off the fat as it rises to the top. Strain, and stir occasionally as it cools to prevent a skin from forming.

Barbeque Sauce
(Makes 5 cups)

In a sauce or stock pot, sauté:
1 medium onion, finely chopped
4 garlic cloves, minced
in:
¼ cup vegetable oil
When the vegetables begin to soften, add:
2 tablespoons vinegar
1½ cups water
½ cup lemon juice
6 tablespoons dark brown sugar
1 cup low-salt tomato sauce or homemade catsup
1 tablespoon dry mustard
⅓ cup celery, finely diced
Cook on a low heat for 35–45 minutes. Season with:
1 teaspoon salt
cayenne pepper to taste
Cool and refrigerate.

Quick White Sauce

(Makes 2 cups)

In a small saucepan, melt:
4 tablespoons butter or margarine
Stir in:
4 tablespoons all-purpose flour
Continue cooking for 5–7 minutes, stirring constantly.
Add slowly:
2 cups milk
Simmer slowly and season with:
salt and pepper to taste
pinch nutmeg
(If necessary, use a wire whisk to smooth out any lumps.)

Basic Fresh Tomato Sauce

(Makes 1½ gallons)

In a large saucepot, heat:
1¼ cups olive oil
Add:
3 cups onions, finely chopped. Cover pot, lower the heat and allow the onions to cook until they are soft and have a little bit of color
Add:
10 pounds ripe tomatoes
8 ounces salt-free tomato paste
1 tablespoon dried basil
2 teaspoons dried oregano
salt and pepper to taste
Simmer for 5 minutes, stirring occasionally. Uncover pot and add:
15 garlic cloves, minced

Stir briefly and add:
8 cups water
Simmer sauce over low heat for 1 to 2 hours. If sauce appears too thick add more water; if too thin add more tomato paste. This can be prepared in advance and frozen in smaller quantities.

Chili Sauce
(Makes 1 quart)

Bring large pot of water to boil. Wash:
4 quarts fresh tomatoes
Make an "X" at the base of each tomato with the tip of a paring knife. Drop the tomatoes into boiling water for 20 to 30 seconds. Remove from water and peel off the skin. Quarter each tomato and place them in a large, heavy pot. Add:
3 green chili peppers, diced
2 large onions, diced
¼ cup brown sugar
1½ cups cider vinegar
1 teaspoon celery seed
½ teaspoon cinnamon
¼ teaspoon ginger
¼ teaspoon nutmeg
1 tablespoon salt
2 teaspoons black pepper
1½ teaspoons allspice
¼ teaspoon cloves
1 teaspoon dry mustard
Simmer very slowly, stirring occasionally, until mixture is very thick (approximately 2 hours). Season with salt to taste. Puree mixture in a blender or food processor to desired consistency. Refrigerate or freeze sauce in small portions.

Worcestershire Sauce
(Makes 2-3 cups)

If you search the health food stores you will probably find a few brands of worcestershire sauce that are MSG-free. But ambitious chefs will want to try this recipe.

2 cups cider vinegar
2½ tablespoons catsup
1 ounce finely chopped anchovies
2 tablespoons chili sauce (recipe above)
pinch cayenne pepper
2 teaspoons sugar

1. Place all ingredients in an airtight container and shake 3–5 times a day for 2 weeks. Then strain, pour into sterilized bottles, and seal.
2. Store in a cool, dry place.

Chicken Stock
(Makes 1½ gallons)

8 lbs. chicken bones, uncooked (if you have cooked chicken bones on hand, you can use them)
5 large onions, quartered
3 large carrots, cut in 1-inch pieces
2 ribs celery, cut in 1-inch pieces
2 bunches parsley stems
6 bay leaves
3 tablespoons dried thyme
20 black peppercorns
2 whole cloves

1. Place rinsed chicken bones in a large stockpot. (Do not rinse cooked bones.) Cover with cold water to 4 inches above the bones. Bring to a boil.
2. Reduce heat and simmer 10 minutes.
3. Skim foam from the top and discard. Add the remaining ingredients and continue to simmer for 4–5 hours.
4. Pass the stock through a fine strainer, gently pressing the vegetables and bones to release more flavor.
5. Discard the solids and cool the liquid to room temperature. Refrigerate overnight and then skim off the fat. Refrigerate or freeze in small portions.

Beef Stock

(Makes 1-1½ gallons)

10 pounds beef bones
⅔ cup vegetable oil
8 large onions, cut in 8 segments
6 large carrots, cut in 1-inch pieces
6 ribs celery, cut in 1-inch pieces
12 ounces low salt tomato paste
1 cup red wine or water
8 bay leaves
20 black peppercorns
½ bunch parsley stems
2 tablespoons thyme

1. Preheat oven to 425 degrees.
2. Place the bones in a large baking pan and coat with vegetable oil.
3. Bake for 1 hour or until the bones begin to brown, occasionally turning them for even browning.

4. Mix in the onions, carrots, and celery. Bake for another 30 minutes and then add the tomato paste.
5. Bake another 30–40 minutes or until the bones and vegetables are well browned but not burnt.
6. Remove from the oven and place the bones and the vegetables in a large stockpot. Place the baking pan over high heat. When the sediments in the bottom of the pan begin to cook, add the wine or water and mix with the sediments. Pour this mixture over the bones in the stockpot.
7. Add cold water to 4 inches above the bones and bring to a boil. Skim the foam off the top of the mixture and reduce the heat to a simmer.
8. Add the thyme, peppercorns, bay leaves, and parsley and continue to cook over low heat for 5–7 hours.
9. Strain the stock through a fine mesh strainer and discard the solids. Allow the stock to cool at room temperature and then freeze or refrigerate.

Brown Stock
(Makes 2 quarts)

Cut in pieces and brown in a 350-degree oven:
6 pounds shin and marrow bones
Place the bones in a large stockpot with:
4 quarts water
8 black peppercorns
6 whole cloves
1 bay leaf
1 teaspoon thyme
3 sprigs parsley
1 large diced carrot
3 diced celery stalks
1 cup drained canned or fresh tomatoes
1 medium diced onion
1 small white, diced turnip

Bring to a boil. Reduce heat and simmer uncovered for 2 to 3 hours or until reduced by half. Strain stock. Cool uncovered and refrigerate.

Quick Herb Stock

(Makes 1 quart)

Place in a soup pot:

2 tablespoons dried thyme
2 tablespoons dried parsley
1 tablespoon dried oregano
10 black peppercorns, crushed
4 bay leaves
1 medium onion, sliced thin
2 quarts cold water

Bring to a boil and simmer until the liquid is reduced by one-half. Pass through a fine strainer and cool. Refrigerate or freeze.

Hot or Cold Tomato Herb Soup

(Makes 18-20 portions)

½ pound butter or margarine
7 cups onion, sliced
8 garlic cloves, minced
1 cup fresh dill
¼ cup fresh thyme, chopped or
  1 teaspoon dried thyme
8 pounds fresh ripe tomatoes or canned plum tomatoes, drained
¼ cup fresh oregano, chopped or
  1 teaspoon dried oregano

⅓ cup fresh basil, chopped or
 1 teaspoon dried basil
salt and pepper to taste
18 cups chicken stock or water
1 teaspoon ground allspice
½ teaspoon ground mace
1 tablespoon sugar

1. Heat butter or margarine in a large stock or soup pot. Add onions and cook over low heat, covered, until soft and translucent.
2. Add garlic and cook for 3–5 minutes. Stir in all remaining herbs.
3. Add the tomatoes, chicken stock, allspice, mace, and sugar. Bring soup to a boil. Reduce heat to medium-low and cover.
4. Cook for 40–50 minutes and remove from heat. Puree soup in a blender, food processor, or food mill until smooth. Reheat and season with salt and pepper.
5. Serve immediately while hot or refrigerate and serve cold with a dollop of yogurt or sour cream.

Red Meat Marinade
(Makes 2 cups)

¼ cup peanut oil
¼ cup olive oil
½ cup red wine vinegar
¾ cup red wine
3 cloves garlic, peeled and cut in half
10 crushed black peppercorns

Whisk ingredients together. Pour over beef or pork and marinate in refrigerator overnight or at room temperature for 3 to 6 hours.

Catsup
(Makes 3 quarts)

In a large stockpot, place:
- 5 quarts tomatoes, cut in pieces
- 5 medium onions, sliced
- 1 red pepper

Simmer until all ingredients are soft. Puree in a food processor, blender, or food mill. Return to pot and bring to a boil. Add:
- ¼ cup sugar
- ¼ cup brown sugar

Place in a cheesecloth bag:
- 1 cinnamon stick
- 1 teaspoon whole cloves
- 2 teaspoons allspice
- 1 teaspoon mace
- ¼ teaspoon nutmeg
- 1 teaspoon dill seed
- 2 teaspoons celery seed
- ¼ teaspoon dry mustard
- ½ garlic clove
- 1 bay leaf

Place the spice bag in the tomato mixture and boil until the liquid is reduced by one-half. Remove the spice bag and add:
- 2 cups apple cider vinegar

Lower the heat and simmer 8–12 minutes. Add to taste:
- salt
- cayenne pepper

Refrigerate or pour into sterile jars and seal.

Mayonnaise
(Makes 2 cups)

Place in blender:
- 1 egg

1 teaspoon ground mustard
1 teaspoon salt
dash of cayenne
1 teaspoon sugar
¼ cup salad oil

Cover and blend until thoroughly combined. With blender still running, slowly add:

½ cup salad oil
3 tablespoons lemon juice

until thoroughly blended. Then add slowly:

½ cup salad oil

and blend until thick.

APPENDIX 2

Account of Our First Observations of MSG Asthma

By Dr. Gary Baker

Our discovery of MSG asthma was both fortuitous and dramatic. It happened in 1981, whilst I was completing my training as a Respiratory Physician with Dr. David Allen at the Royal North Shore Hospital, Sydney, Australia. We were conducting a study into the clinical importance of a number of food additives as precipitants of acute asthma. The additives that we had initially selected to study were metabisulphite, tartrazine, salicylate and benzoates. MSG was not included in the early studies as we did not expect it to be a precipitant.

A number of people suffering from asthma that was severe enough to cause them to be admitted to hospital agreed to participate in the study. Essentially, they agreed to adhere to a strict diet, which excluded the above food additives, for two weeks, and then to ingest a capsule containing either a food additive or sugar, and to be observed for 6 to 12 hours with repeated tests of air-flow rates, which is an objective test for asthma.

One patient with chronic asthma, who was found to be sensitive to a number of the food additives, had a marked improvement in controlling her asthma, when she continued to exclude those particular additives from her diet. However, one evening she broke her diet by having a Chinese meal at a restaurant. During the night her asthma was similar to previous nights, but, in the morning, 10 to 12 hours following the meal, she became increasingly breathless despite using her medications. She went to hospital and was admitted for intensive inhalational and intravenous therapy. Her asthma attack was unresponsive to even this treatment and continued to worsen. She was placed on a ventilator and required cardiac support. Thankfully, her asthma finally relented, she recovered and was able to leave hospital.

We were very concerned over this life-threatening episode, which seemed to be related to the ingestion of the Chinese meal. Although the meal may have contained additives to which this patient was known to be sensitive, the severity of her reaction raised the question whether there were additional substances in the Chinese meal to which she was reacting. This is when we first considered the possibility that the addition of MSG to food may be a precipitant for asthma in particular people.

This possibility was strengthened by our observations when the patient agreed to take MSG according to our protocol for other food additive testing. A capsule containing 2.5 grams of MSG, which is an amount that can be ingested during a Chinese meal, was administered. No other additives were ingested and the patient was on her food additive exclusion diet. There was no change in her ability to breathe until after approximately 12 hours. The asthma quickly became severe and, again, was relatively unresponsive to normal hospital asthma therapy and required ventilatory support. Thus, this challenge appeared to reproduce both the severity and the delay that had been seen following her earlier experience with the Chinese meal.

The severity of this asthma attack following the MSG challenge surprised and worried us. When another patient, whom we were currently investigating for food additive asthma, also reacted to a MSG challenge with a marked decrease in her airflow rate, we wrote letters to alert the medical community of the potential dangers of MSG in asthmatic patients (*New England Medical Journal*, 1981, 305: 1154–1155; *The Medical Journal of Australia*, 1981, 2: 576). Dr. Jack Delohery joined us to undertake a larger clinical study. These studies have supported and further characterized our initial observations of MSG as a precipitant of asthma in some asthmatics, and have been reported in the *Journal of Allergy and Clinical Immunology*, 1987, 80: 530–537.

INDEX

Abdominal pain
  stomach cramps, 13, 14
  severe upper abdominal pain, 15
  stomachache, 22
  stomach pain and asthma, 24
  stomach pain, 25
Abdominal swelling, 15
Accent salt, 16
Acetylcholine, 28
  factor in depression, 3
"Agent Blue," 3
Aging, MSG effects, 32
Airplane food, 87–88
Aji-no-moto Company, 1
  challenged by Chinese, 6
  one-half of world supply producer, 6
"Aji-no-moto," or ajinomoto, 6, 53
Albert Einstein School of Medicine, studies on MSG, 11–12
Aliases of MSG, 53
"All natural," containing hydrolyzed vegetable protein, 31
Allen, Dr. David, 23, 109–110
Allergic reaction, differentiation from toxic reaction, 36–37
Allergic symptoms, 39, 41
Altered reality, 91
"Altitude rash," 19
Alzheimer's disease, 33, 45
American eating behaviors, 36
Amino Products Corporation, 8
Amtrak, use of MSG, 87
Amyotrophic Lateral Sclerosis, 33, 45
Andrews, John, 9
Anhedonia, 26. *See also* Depression
Animal studies of MSG, 44
Anson, Dr., 9
Anti-freeze, MSG solution used as, 7
Arby's, 83

Archives of Neurology, The, Greenamyre study, 33
Armed forces, sponsor of first MSG symposium, 8
Arthritis-like symptoms, 17–19
Asthma
  associated with MSG, 2
  experimental studies, 23–25
  historical account of association with MSG, 109–110
  increase in children, 25
  increased death rate, 25
Atopic dermatitis, 19
Australia, case reports from, 38
Avoiding MSG, 48

Baby foods
  MSG removed from, 3, 21
  soups fed to babies, 63
Bacon, 35
Bacon bits (artificial), 60
Bacon, Sir Francis, 89
Baker, Dr. Gary, 23, 38, 45, 109–110
Baking supplies, 52
Balance difficulties, 32, 33
Barbecue sauce (recipe), 99
Basic Fresh Tomato Sauce (recipe), 101
Beef Gravy (recipe), 97
Beef Stock (recipe), 103
Behavioral disturbances in children, 14, 35
Boller, Anna, 9
Borden, represented at initial symposium, 9
Borden, Gail, role in canning industry, 90
Bouillon, 35, 36
Bowel control, 14
Bowers, Mr., 9
Box dinner, 51
Bloom, definition, MSG enhances, 43. *See also* MSG, as flavor enhancer
Brain, 43
Breads, 52
Breathing difficulties, 37
Broths, 48, 62, 64
Brown Gravy (recipe), 102–103
Brown Stock (recipe), 104–105
Burger King, 83
Burning sensation, 13, 15
Byck, Dr. Robert, 38

Cafeterias, 86–87
Caffeine, 37
Campbell Foods, represented at initial symposium, 9
Canned foods
  containing MSG, substitutions, 48–49
  history of, 90
  increased use by families, 35–36
  listing of supermarket foods, 72, 74
  low sodium, 50, 64
  most likely to contain MSG, 51, 52, 66
  older people rely on, 32
  tinny taste removed, 10
Case histories, 16–32

Catsup, 49
Catsup (recipe), 107
Central (brain) symptoms, 39
Cereals, 52
"Challenge dose" of MSG, 24
"Chase and Grabbits," 35
Chest tightening and pressure, 15, 30, 85
Chicken soup with rice, 63
Chicken Stock (recipe), 102
Children
- asthma, 25
- behavioral reactions, 3, 13–14
- depression, 27
- hyperactivity, viii, 13
- learning disabilities, case report Noah, 13–14
- shiver-shudder symptoms, 21–22
- *See also* MSG reaction in children and teens

Chili, canned, 48
Chili Sauce (recipe), 101
Chinese, and production of MSG, 6–7
"Chinese Restaurant Syndrome"
- acetylcholine and, 28
- increased on empty stomach, 42
- initial report, 11
- restaurants, 85–86
- symptoms, 16
- *See also* Oriental foods

Chinese seasoning, 53
Colman, Dr. Arthur D.
- case reports, 13
- depression, 25–26
- foreword by, viii–x

Consumption, 38
Continental Can Company, 9
Continental Foods, 9
Cookies, 51, 75
Crackers, 68, 75
Croutons, 60, 79

Dairy products, 51, 60
Dairy Queen/Brazier, 85
Dalphin, Charlotte, 9
Degenerative diseases (nerve cell), 33 (*See also* names of specific diseases)
Delicatessen food, 75–76
Delohery, Dr. John, 23, 109–110
Denny's, 84
Depression, viii, 2, 13, 20, 25–30, 35, 91
- among children and teens, 3, 27

Diamond Headache Clinic, 31
Diarrhea
- case reports, 14, 16, 18, 20, 29
- in children, caused by MSG, 13

Diet foods, 51, 76
Discrediting of scientists, 45
Disodium guanylate, 55
Disodium inosinate, 55
Dips, 68, 69
Dizziness, 13, 16, 29
Dose-related effects of MSG, 37
Dove, Franklin, 9
Dried foods, 32, 90
Drug effect of MSG, 4, 90, 91
D'Sinter, Mr., 9

Eating patterns, 35, 36
Economy, MSG-dependent, 3
Emotional symptoms, 2, 13–14, 16, 25–35, 41, 91
Encyclopaedia Britannica, MSG in tobacco reference, 36
England, refrigerators in, 89
Epidemic, sensitivity, 35
Epilepsy-like symptoms, 22
Eshel, Dr. Elyakim, 15
Esophagus, 91
"Essence of taste" (*See* Aji-no-moto)
Eyes
  blurred vision, 17, 20
  burning, 13
  flashing lights, 16
  pressure behind, 25

Facial pressure, 13, 21
Facial tingling, 35
Faintness, 20
Fast foods, 4
  doses from, 37
  listing of restaurants, 82
  new category of people, 35, 36
  symptoms after, 13, 16, 31–32
Fast food restaurant guide, 82–85
Fatigue, 13, 20, 29, 93
Federation of the American Society for Experimental Biology, The, 2, 13, 30
Fish, canned, 48 (*See also* Tuna)
Fish, fresh, 52 (*See also* Seafood)
Flavorings, reading labels, 48, 54, 58, 71
"Flavor packets," 48, 55
Flushing of skin, viii, 2, 15, 21, 25, 30
Food and Drug Administration, 52
Food industry, 8–10, 36, 89–90, 91
Food labels (*See* Labels)
Food Power, 93
Food Preparation Network, 50
France, 38, 87
Freeze dried foods, 76
Freezing, 89–90
Frozen foods, 35, 36, 47, 51, 62, 76–78
Fried chicken, 35, 85
Fruits, canned, 52

Gann, Dr. Dietmar, 15
Garlic salt, 61
Gastrointestinal symptoms, viii, 39, 40, 41
Gefilte fish, 35
General Foods, 9
Generally Recognized As Safe (GRAS) substances, hearings and testimony, 30
Germany
  case reports from, 38
  chemical identification of glutamic acid, 5
  source of potash, 7
Gerstl, Dr. Robert, 38
Glutacyl, 54
Glutamate Association, vii, ix
Glutamate neurotoxicity, 33
Glutamic acid, 5, 6, 42
Gourmet powder, 53
GRAS substances, 30, 52
Gravy, canned, 48

Gravy mixes, 78
Greenamyre, Dr. J. Timothy, 33
Guest, Mr., 9

Harassment
 associated with questioning safety of MSG, ix
 of researchers, vii–ix, 45
 See also MSG research
Headaches, vii, 2, 13, 16, 17, 20, 22, 25, 29, 30, 31–32, 35, 93
Health food stores, 47, 57
Heart irregularities, 2, 15 (*See also* Palpitations)
Hiding MSG, 30
History of MSG, 5
Hives, 20
"Homemade," 56
Hospital food, 16
Hot dogs, 67
"Hot flashes," 16
Hot or Cold Tomato Herb Soup (recipe), 106
Huntington's Chorea, 45
Huntington's disease, 33
Hydrolyzed vegetable protein
 associations with Alzheimer's, Parkinson's, ALS, 33
 chemical method to produce MSG, 1–2
 health foods using, 47
 in aging, causing symptoms, 32
 in labels, 55, 61, 62, 63, 64, 66, 67, 69, 71
 in restaurants, 81
 labelling as natural flavoring, 30–31, 48, 52, 53, 54
 MSG found in, 17
 quantities used, 38
 relationship to MSG, 2
 shopping at supermarket, 72
 widespread nature, 35
Hydrolyzed plant protein, 54, 58
Hydrolyzed protein, 59
Hyperactivity in children, viii, 13
Hypothalamus, damage to, 44

Ice cream, 52
Ikeda, Kikunae, 1, 5, 6, 8
Incontinence, 13, 14
Ingelfinger, Dr. Franz, 12
Insomnia, 13
International foods, 51 (*See also* Oriental foods)
International Minerals and Chemicals Company, 8
Intravenous effects of MSG, 12
Irritability, 2
Israel, 38
Israel Journal of Medical Sciences, 15
Itching, 37

Jack in the Box, 82–83
Japan, 1, 8, 89
 growing awareness of MSG syndrome in, 38
 invasion of China, 6–7
Jobe, Inez, 9
Johns Hopkins University, 8
Joint aching, 18–19
Journal of Allergy and Clinical Immunology, 110
 Allen, Delohery, Baker study, 23
Juices, frozen, 52

Kentucky Fried Chicken, 3, 14, 85
Kline, Dr. Nathan, 26, 28
Kombu, 1, 5, 32, 47, 54
Kosher foods, 67
Kwok, Dr. Ho Man, 2, 11
"Kwok's disease," 12

Labelling by FDA regulations, 52–53
Labels
  aliases of MSG, 53
  ambiguities of requirements, 51–53
  hiding MSG, 30–31
  reading labels, 48, 51–69
Laminaria japonica, 5
Larrowe, James E., 7, 8
Larrowe Milling Company, 7
Larrowe-Suzuki Company, 8
Lawrence, Charles, 9
Lawry Company, 61
Libby Food Company, 9
Learning disabilities, case report Noah, 13–14
Lethargy, 19
Light sensitivity, 16
Lind, James, 38
Locking jaw, 21
Logan, Col. Paul, 81
Long-term effects, 45
Long John Silver's, 84
Lou Gehrig's disease, 33, 45
Low-salt foods, 49–50, 87
Lucas, Dr. J. P., 44

Manufacture of MSG
  American, 7–8
  Japanese, 6–7
Marriott Catering Services, 87
Marshall, Albert, 8
Mashman, Dr. Jan, 38
Mason City, Iowa, 7, 8
Matzo balls, 35
Mayer, Dr. Jean, 3
Mayonnaise (recipe), 107–108
McCollum, Dr. Elmer, 8
McDonald's, 83
Meats, 10, 18
  fresh, 52
  canned, 49, 51, 72
  cured, 51, 75
Meat tenderizers, 35
Mechanisms of MSG toxicity, 42, 44, 45
Medical Journal of Australia, The, 110
Mellon Institute of Industrial Research, 7
Menopause-like symptoms, 16, 33
Menstrual cycles, 33
MI (*See* Myocardial infarction)
Migraine headaches, 16, 22
Monopotassium glutamate, 72
Monosodium glutamate (*See* MSG)
Mood swings, 2, 16, 20, 28
Mt. Sinai Medical Center, 15
MSG, vii (*see also* Ajinomoto)
  and armed services, 9–10
  and sweets, 10
  and children, 3, 45, 58, 63 (*See also* MSG reaction in children and teens)
  aliases for, 53–54
  and human diseases, 33 (*See also* names of specific diseases)

animal studies of, 44–46
as drug, 4, 90–92
as flavor enhancer, 4, 36, 43–44, 53
as neurotransmitter, 44
as Oriental food secret, 9
as spice, 1
asthma and, 109–110
avoidance of, 47–50, 51–80, 81–88, 97–108 (*See also* MSG reaction, avoidance leads to cessation of symptoms)
conference about (1948), 9–10, 89–90
consumption increasing, 4, 36, 37, 38
cooking without, 47–50 (*See also* Recipes)
developer of, 1 (*See also* Ikeda, Kikunae)
economic significance of, 91
effects of, vii (*See also* MSG reaction)
flavor of, 43
foods containing, 4, 18, 32, 35, 36, 48–50 (*See also* specific types of food)
history of, 5–10, 89–90
in fast foods, 3 (*See also* Fast foods)
in supermarket foods, 51–80
in U.S. diet, 4, 10, 22, 32, 35–36
Japanese production during 1930s, 6
long-term ingestion, 32–33, 44–45
maintains color in canned foods, 36
neurotoxicity of, 32–33, 45
original substance encased in monument, 1
patented by Ikeda, 6
physiological and biochemical effects of, 90–92 (*See also* MSG reaction; MSG, neurotoxicity of)
recipes without, 49–50, 97–108
seaweed source of, 1, 5, 6 (*See also* Kombu)
substitutions for, 49 (*See also* Substitutions)
staple, 2
sugar beet source of, 7–8
U.S. annual consumption, 38
uses of, 1, 91
why toxic, 42
MSG reaction, 11–33, 81–82, 90–92
and age, 32
and asthma, 23, 25, 37, 109
avoidance leads to symptom cessation, 14, 15, 18, 19, 21, 22, 24–25, 26, 29, 32, 109–110
biochemical mechanism of, 28
can lead to suicide, 2
can provoke asthma attacks, 25, 109–110
case histories, viii, 2, 3, 13–33
debilitating and/or life-threatening, 2, 4, 14, 15, 17–19, 24, 25, 26, 28, 29, 31, 45, 109–110
emotional symptoms, 2, 13–14, 16, 25–35, 41, 91

MSG reaction (continued)
growing awareness of, 38
in children and teens, viii, 3, 13–14, 21–22, 30, 35, 40–41, 92
lack of awareness about, 27, 38
mistaken for heart attack, 12
not only reaction to Oriental food, 4
not usually allergic reaction, 4, 36–37
portion of population affected, 21–22, 27
range of, 2, 15
symptoms of, viii, 2, 4, 27–28, 29, 30, 32–33, 39–42 (*See also* Symptoms; names of specific symptoms)
to elevated dosages, 30, 32–33
to small dosages, 30 (*See also* MSG sensitivity)
MSG research, vii, 11–15, 21, 27, 29–30, 35, 38
and asthma, 23–25, 109–110
harassment of researchers, viii–ix, 45
too diffuse focus, ix
MSG sensitivity, vii, ix, 2–3, 31, 44, 92
many afflicted are unaware, 2
of epidemic proportions, 35
portion of population affected by, vii, 2–3, 13, 21–22, 27
MSG syndrome. *See* MSG reaction
MSG toxicity, ix, 4, 12
at experimental level, 29–30, 44
Muscle soreness, 17–19
Myocardial infarction, 12

National Institute of Allergy and Infectious Diseases, 25
National Institute of Mental Health, concern about depression, 3
National Livestock and Meat Board, 9
National Press Club, MSG removal from baby food, 3
National Restaurant Association, 81
National School Lunch program, 86
"Natural flavorings"
may include hydrolyzed vegetable protein, 31
reading labels, 48, 60, 62, 65, 66, 67
food labelling requirements (FDA), 52–55
Natural flavors, 17
Natural foods, 47
"Natural ingredients," 47
Nausea, 13, 16, 25, 35
Nestles, 9
Neurotransmitter, glutamate as, 44
New England Journal of Medicine, viii, 12
Allen and Baker study, 23
Colman's letter, viii–ix

Ho Man Kwok's study, 2, 11
Reif-Lehrer study, 22
Newhouse, Dr. D., 44
Noah, case report, 13–14
"No Artificial Ingredients," 65
"No Preservatives," 65
NPD Group, study of meal preparation, 36
Numbness, 11, 15

Olney, Dr. John, 44
Oriental food, 18, 19, 20, 21, 22, 28, 31, 85. *See also* "Chinese Restaurant Syndrome"
Oscar Mayer, 9

PMS. *See* Pre-menstrual syndrome
Palpitations, 15
Pancake mixes, 52
Paranoia, 2, 26, 41
Parkinson's disease, 33
Pasta, 52
Patent, original manufacture, 6
Peripheral responses, 39
Pfeiffer, Dr. Carl, experiments to raise IQ levels with MSG, 10
Pillsbury, 9
study of eating behaviors 1971–86, 35–36
Pizza, carry out, 36
Pizza Hut, 84
Poison, MSG as, ix, 2
Potash, 7
Potato chips, 51, 68, 79
Poultry Gravy (recipe), 98
Pre-menstrual syndrome, 33
Prepared foods, 72–74
Processed foods, 32, 37, 89
Pre-packaged foods, 26, 32, 47
Pressure around eyes, 35
Protein Hydrolysate, labelling as natural flavoring, 53
Pullium, H. H., 86

Quick Herb Stock (recipe), 105
Quick White Sauce (recipe), 100

Rations (Army), 8
Ratner, Dr. David, 15
Rash, 19, 37
Recipes, 49–50, 97–108
Red Meat Marinade (recipe), 106
Refrigerators, 89
Reif-Lehrer, Dr. Liane, 13, 21, 27
GRAS testimony, 30
Research. *See* MSG research
Restaurants
airplanes, 87
Arby's, 83
Burger King, 83
cafeterias, 85
Dairy Queen/Brazier, 85
Denny's, 84
Jack in the Box, 82
Kentucky Fried Chicken, 85
Long John Silver's, 84

Restaurants (continued)
McDonald's, 83
Pizza Hut, 84
Schools, 86
Taco Bell, 84
trains, 88
Wendy's, 82
Retinal degeneration, 44
Rittenhausen, 5
RL-50, 54
Royal North Shore Hospital (Sydney, Australia), 23, 109–110

Salad dressings, 18, 48, 51, 60, 79, 87
Salads, prepared, 60, 70, 79
Sauce mixes, 56
Sauces, 18, 35, 48, 49, 80, 86
Sausages, 35
Seafood, 10, 87–88. *See also* Fish
Seasonings, 18, 35, 78, 80, 86
Sea tangle, 1, 5. *See also* Kombu
Seizures, 22
Senility, 33
Schaumberg, Dr. Herbert, 11–13, 38
School lunches, 21, 86–87
Science, 12–13
Schaumberg, Byck, Gerstl, and Marshman report, 38
Scurvy, 38
Shaking, 20
Shivers, 22
Shopping
tips to avoid MSG, 48
substitutions, 49
shopping smart, 72–80
Shortness of breath, 13, 15. *See also* Asthma
Shoshani, Dr. Ehud, 15
"Shudder" attacks, 21–22
Slurred speech, 20
Smithson, Linda, 36
Snack foods, 68, 79, 88
Sneezing, 20
"Sodium free," 72
Soft drinks, 52
Soup mixes, 18
Soup stocks, 36, 49, 65
Soups, 1, 20, 22, 23, 32, 63, 64, 69
canned, 36, 49, 51, 66, 74
dried, 48, 51, 74
Southern Medical Journal, 15
Soy sauce, 1, 36, 49–50, 86
Spaghetti and meat balls, 35, 58
Spices, 80
"Spices," 48, 66
Standard Brands, 9
"Steffen's waste water," 7–8
Stew, 10
canned, 48
Stokeley, 9
Stomachaches, 13, 14, 15, 16, 22, 24, 25
Stroke-like symptoms, 20
Substitutions, 49–50
Subu, 53
Sugar beets, 7
Suicide, 10, 27
Sulfites, 70
Suzuki, Saburosuke, 6, 8
Suzuki Spice Company, 7
Sweating, 15
Swelling, 15, 20, 37

Symposium, 1948, 8
  marking American revolution in food, 8–9
Symptom analysis, 39–41
Symptoms
  allergic symptoms, 41
  central (brain), 41
  chest, 39
  children's symptoms, 40–41
  eye, 40
  gastrointestinal, 39
  list of reported symptoms, 40
  peripheral, 41
  symptom analysis, 39

Taco Bell, 84
Tamari sauce, 49
Taste, altered, 43
  flavor enhancer, 43
Taste buds, 43, 44
Tendonitis, 17, 18
The Value of Glutamate in Processed Foods, 81
Tight sensations, 25
Tingling, 85
Tinny taste, reduced by MSG, 90
Tobacco, cured with MSG, 36
Tokyo University, 1, 5, 6
Toxic reaction, 36–37, 42 (*See also* MSG, neurotoxicity of)
Train food, 87–88
Tressler, Dr. Donald, 7
Tuna fish, 59, 70

Unami, 6
United Airlines Food Service, 9
United States Department of Agriculture, 86
University of Michigan Neuroscience Laboratory, 33

Vegetables
  canned, 52
  frozen, 52
Vegetable powder, 57
"Vegetable protein," 48 (*See also* Hydrolyzed vegetable protein)
"Vegetable broth," 59
Ventricular tachycardia, 15
Ve-tsin, 6, 53
Visual disturbance, 13, 16 (*See also* Eyes)
Vomiting, 20, 22
Volper, Cindy (Chef–Recipes), 97

Wall Street Journal, The
  Asthma death rate article, 25
  Pillsbury study article, 36
Washington University (St. Louis), 44
Waste water, 7–8
Weakness, 13, 15, 20, 29
Wendy's, 82
Wolff, 5
Wonton soup, 37, 95
Worcestershire Sauce (recipe), 102
World War I, 7
World War II, 8

ABOUT THE AUTHOR

George R. Schwartz, M.D., an internationally known physician and toxicologist, is the author of *Food Power* (McGraw-Hill 1979) and senior editor of the widely acclaimed textbook *Principles and Practice of Emergency Medicine* (W. B. Saunders Co. 1978, 2nd ed. 1986).

A native of Caribou, Maine, Dr. Schwartz did his undergraduate work at Hobart College and then attained his M.D. degree Magna Cum Laude from The State University of New York, Downstate Medical Center. He received internship and residency training in Seattle, Washington, Indiana and New York City.

He has served on the faculty of two medical schools, and currently is in clinical practice in Santa Fe and Los Alamos, New Mexico.

The father of five children, Dr. Schwartz lives with his family in the mountains near Santa Fe, New Mexico.